Survival Strategies for
Agricultural Cooperatives

Survival Strategies for

Agricultural Cooperatives

CHARLES E. FRENCH
JOHN C. MOORE
CHARLES A. KRAENZLE
KENNETH F. HARLING

Iowa State University Press, Ames, Iowa

To JEANNE, MEG, KATHERINE, and CATHY

Charles E. French is Study Director, Food and Nutrition, President's Reorganization Project, Executive Office of the President, Washington, D.C., and holds the M.A. degree from the University of Missouri and the Ph.D. degree from Purdue University.

John C. Moore is Assistant Director, Economic Analysis Division, Office of Administration, Farm Credit Administration, Washington, D.C., and holds the M.S. and Ph.D. degrees from the University of California at Berkeley.

Charles A. Kraenzle is Agricultural Economist; Economics, Statistics and Cooperatives Service, USDA; and holds the M.S. degree from the University of Missouri and the Ph.D. degree from the University of Connecticut.

Kenneth F. Harling is Research Assistant, Department of Agricultural Economics, Purdue University, holds the M.S. degree from Purdue University, and is a candidate for the Ph.D. degree from Purdue University.

Composed and printed by The Iowa State University Press, Ames, Iowa 50010

First edition, 1980

Library of Congress Cataloging in Publication Data
Main entry under title:

Survival strategies for agricultural cooperatives.

Includes bibliographical references and index.
1. Agriculture, Cooperative—United States. I. French, Charles Ezra, 1923-
HD1491.U5S95   334´.683´0973   79-20528
ISBN 0-8138-0455-8

CONTENTS

FOREWORD

Who will control United States agriculture? Who will market your products? These questions are being asked of and by American farmers. Several options exist. Some suggest that the agriculture of the future will be composed of integrated production marketing systems like the broiler and turkey industry. Others suggest that agricultural markets using the computer as an auction pricing mechanism will replace today's antiquated livestock pricing systems. Still others see bargaining between farmers and processors as the prototype for the future.

The exact outcome is not clear. More likely than not, what evolves will be a combination of systems.

What is clear is that the strategies pursued by producer-owned cooperatives will have substantial impact on the outcome. Producers acting as individuals can have little impact. Producers acting in groups engaged in farm supply and marketing activities can have substantial impact.

Some, however, question whether today's cooperatives are in a position to compete in the agriculture of the 1980s and 1990s. It is from time to time suggested that they have lost touch with their members, that they are not sufficiently innovative or basic in production and marketing technology, that their management lacks vision, and that they do not have the resources.

Even if the questions are answered in support of the cooperative system, questions exist as to the adequacy of today's government policies and programs affecting cooperatives. On the other hand, more questions are being raised by society and government regulatory agencies as to whether present cooperative tax and antitrust exemptions are in the public interest.

It was in this environment that the USDA encouraged and supported a study of cooperatives' competitive position in United States agriculture. As project leader it sought an individual who had spent his lifetime working as a university researcher and consultant for both cooperative and particularly corporate agribusiness. Charles E. French, formerly Professor and Head of the Department of Agricultural Economics at Purdue University and currently Director, Food and Nutrition Study, President's Reorganization Project, filled this need admirably. Next, it needed two bright young economists who were both willing to ask the hard questions and able to conceptualize what could be. John C. Moore and Charles A. Kraenzle, both Farmer Cooperative Service, USDA, employees at the time the study was done, filled this need. Most good studies need a bright, hardworking graduate student to keep the others honest and to get much of the work done. Kenneth F. Harling, Purdue University, filled this need.

Any cooperative manager, director, or leading farmer who expects to be in the business of agricultural cooperatives during the remainder of the twentieth century should read this book. Agribusiness corporate management and investors should find that this book provides a useful evaluation of their present and future competition. Marketing and business researchers or teachers will find the book a useful guide to cooperative strategies and institutions, with a wealth of research results and ideas for further exploration.

Ronald D. Knutson

Professor and Extension Economist
Agricultural Policy and Marketing
Texas A&M University
College Station, Texas

PREFACE

THIS BOOK was written because of the concern of agricultural leaders about the direction cooperatives are taking. Most cooperative managers, directors, farmers, and others seldom have the opportunity to view the total picture. Too often decisions are guided by history or current specific problems.

This book not only brings together current and future issues in a total framework but also presents alternative strategies for farmers and cooperatives in light of the expected future environment.

The book also has the following dimensions:

• A collection of important and relevant knowledge available on cooperatives.

• A report of the judgment and forward thinking of recognized leaders directing successful United States cooperatives.

• An opinion on the needs of cooperatives by a selected group of academic thinkers.

• A view of the general environment in which cooperatives will function.

• An orientation for much needed longer-run, strategic planning by cooperatives.

• A summary containing implications drawn from the above agricultural cooperative managers, cooperative leaders and thinkers, and policymakers as viewed by four professionals who had a unique opportunity to concentrate for an extended period on the needs relevant to such a book.

Information for this book was obtained from a number of sources. The most important source was personal discussions by the authors with nearly 150 cooperative managers and leaders throughout the United States. Information by questionnaire was also obtained from nine cooperatives (referred to as the "core sample") and nine academic people. An extensive array of publications and articles on cooperatives and related material was reviewed.

The approach presents an overview and analysis of a wide range of subject areas. Specific commodity or regional areas are not discussed separately. Emphasis is given to agricultural cooperative functions and strategies as they could be applied to cooperatives generally.

The authors extend their appreciation to all who assisted, especially the cooperative leaders and academic people. Without their sincere interest in the furtherance of cooperation and their candid answers, this book would not have been possible.

The authors gratefully acknowledge the many comments and suggestions received from the following people who reviewed all or parts of the book: J. Warren Mather, Jack Armstrong, Dave Volkin, Nelda Griffin, James Haskell, George Tucker, James Baarda, J. David Morrissy, Clement Ward, Donald E. Hirsch, and Donald Vogelsang, ESCS; Ronald Knutson, Texas A&M University; and Lee Schrader and Emerson Babb, Purdue University.

Appreciation is also extended to Purdue graduate students Paul Blick and Rod Paschal for their assistance, typists Patricia Burns and Loraine Hill, astrocomp operators Debbie Thomas and Debbie Martin, and many others who helped in preparing the manuscript.

The authors also extend their appreciation to Margaret Moore, Katherine Kraenzle, and Catherine Harling for their sacrifices and support during the preparation of this book.

Survival Strategies for
Agricultural Cooperatives

Good pasture, beef cattle, adequate water supply, and a well-kept farmstead give a picture of prosperity.

1 General Environment for Cooperatives

AGRICULTURAL COOPERATIVES have helped farmers adjust to changing times and conditions in the United States since the mid-1800s. They have done this well and have grown into a vital agricultural institution. But helping farmers adjust to their environment was not easy, and it will be even more difficult in the future.

The environment that cooperatives[1] face for the rest of this century is expected to be so formidable that some doubt exists whether cooperatives will survive. Many cooperative leaders do not accept this doubt. They are pleased to meet the challenge of the future.

Certain strategies can help cooperatives to lead farmers in adjusting to their environment and also help farmers actually to change the environment. Consequently, cooperatives influence the environment as well as being influenced by it. So the notion of cooperative survival is up to the cooperatives themselves.

Historically, agricultural cooperatives have helped farmers face their environment in three important ways: providing competitive outlets through which farmers can be assured of marketing channels; providing farmers with a competitive source of supplies; and providing farmers a competitive basis for acquiring services such as credit and electricity. In some instances, cooperatives have benefited farmers by providing a voice to obtain desired legislation.

Cooperative members have been more directly helped by cooperatives than have nonmembers, but the fallout advantages for all farmers have been important.

Cooperatives can still do much for agriculture by continuing these traditional activities, but the changing environment will demand much more of cooperatives. They must provide countervailing power of a strong competitive relationship to other forms of agricultural businesses and provide a

3

viable alternative to the current trend toward corporate (noncooperative) intervention into agriculture. Consideration should be given to the type of planning needed; the type of board-management teams that are evolving; the multi-cooperative organizations being developed; the new international horizons for cooperatives; new leadership roles of cooperatives in protecting the individual farm system of production; the need for cooperatives to stand up and be counted on issues such as threats to the Capper-Volstead Act; threats to cooperatives from internal squabbling among different types of cooperatives; and many other things likely to be spawned by the changing environment.

Cooperatives do not have a well-designed strategy nor an information source needed for the future, but some important alternative strategies and information sources are evolving.

The future of cooperatives is not important just to farmers and cooperative leaders. Others such as policy planners, government agencies, and those interacting with cooperatives have a stake in cooperative survival.

Many studies have been done to help cooperatives and several books about cooperatives have been written, but none of these studies or books has focused on alternative strategies that will help cooperatives and farmers to adjust to their environment.

Cooperative leaders must think not only about the general environment but also about the special parts of their environment such as those affected by science and technology, public policy, social trends, and economics. Such thinking is necessary not only to adjust but also to survive.

The general environment for cooperatives is complicated, intertwined, and changing. No one can describe it accurately at any one time because inevitably it will change while one is trying to describe it. Likewise, that very situation is the core of the dilemma faced by the strategist in an agricultural cooperative. This chapter is designed to give that strategist a few handles with which to grab hold of these changing times. The ideas are meant to be illustrative, provocative, and suggestive. This book provides the type of data to guide the reader, but he must adapt and modify it with up-to-date and unique information about the environment faced by his cooperative.

Science and Technology

Scientific undergirding, both public and private, is an acknowledged factor in the success of American agriculture. Former Undersecretary of Agriculture J. Phil Campbell said: "There are good reasons American agriculture leads the world. First, we inherited and nurture a politico-economic system that encourages farmers to produce. . . . Secondly, our American agricultural miracle springs from a unique system of scientific discovery and information dissemination that invariably meets and usually surpasses the needs of its time."[2]

Public support of research and education was built into the environment early: the USDA was created in 1862; the Morrill Act created the land-grant university system the same year; the first state agricultural experiment station was opened in 1875; the Hatch Act, providing a national research network, was passed in 1887; and the Smith-Lever Act created the Cooperative Extension Service in 1914. The private sector research programs were built in a complementary way soon thereafter, and estimates now place the public and private sector research efforts at about the same size.

Public support for research was oriented primarily towards farmers, but cooperatives sought and received much public support for themselves. The need for such research was formally acknowledged by the Cooperative Marketing Act of 1926.

Agricultural firms received increased public support under the Agricultural Marketing Research Act of 1946. By this time noncooperative firms had established their own internalized system of research. In many instances these were based on the specific commodity orientation of the firm. Thus farmers also benefited from it. Public support of cooperative research emphasized the fact that cooperatives lagged behind their noncooperative counterparts, although some cooperatives had started their own research programs as many as thirty-five years earlier.[3]

A Farmer Cooperative Service study in 1973 reported substantially increased emphasis on research by cooperatives—fifty-five cooperatives spent $10.5 million on research in 1972.[4] Increased emphasis on research and development by cooperatives has been due to various factors. One was the de-emphasis on commodity-specific research in the old-line, specialized food companies as they conglomerated. Another factor was that public research emphasizing larger noncooperative firms became more difficult to justify. Also, research in noncooperative firms was often secret. Greater research emphasis on nonfarm food potentials by noncooperative firms also stimulated cooperatives' research. Recently, cooperatives have been evaluating the appropriateness of their research programs, and some plan expansions of their programs.

Food is currently being emphasized as a strategic factor in the survival of the human race. Many studies have emphasized both the domestic and the world food issue, along with the importance of research in this effort.[5] These studies emphasize the issues of worldwide research organizations and the importance of strategic planning to avoid mistakes such as those so vivid in the inadequate United States research program for energy.

A current issue involves appropriate division of research effort between the private and public sectors. Another issue is the organizational arrangement under which research will be conducted. Another is the proportional division of emphasis between basic research and applied research.

Expected developments in United States food and nutrition research include (1) reduced emphasis on food and nutrition research relative to other research in United States food conglomerates; (2) more emphasis on research with worldwide implications; (3) more emphasis on crop-related research relative to livestock-related research; (4) more emphasis on nonfarm, raw-material food relative to farm-based foods, especially in product development and promotion; (5) more emphais on basic research relative to applied research; (6) more concern for evolving technologies to meet food and nutrition needs and not so much concern about the impact of any one technology on historical institutions and precedents; (7) more emphasis on public food and nutrition research as compared to private research; (8) a broadening of United States institutional involvement in food and nutrition research with proportionately less emphasis on the traditional USDA–land-grant establishment; (9) more research findings available from other countries; (10) more emphasis on nutrition and health aspects of food; and (11) more research on food regulations.

Various studies have tried to establish food and nutrition priority research areas.[6] One broad study involving scientists, farm leaders, business leaders, and others included the following priorities: energy research applied to agriculture, with particular regard to the food processing and distribution area which uses much more energy than does farm production; soybean production research; water use and efficiency research; basic research on plant growth and production; and research to determine nutritional requirements of people.[7]

Energy and mechanical research to ease the stress of shortages and their impact on farmers and cooperatives must be increased. Travis W. Manning[8] develops this need effectively.

> In the period following World War II, the notion that we had at last achieved the power, the knowledge, and the resources to control our social and economic destinies became popular. Many people accepted as articles of faith that economic growth would cure most social ills and that scientific and technological capabilities and the supply of resources were adequate for indefinitely sustained economic growth. The first article of faith was shaken by the persistence of poverty amidst plenty at home and the Malthusian specter of exponential population growth and recurrent famine abroad. The second article of faith was weakened by the increasing awareness of the wastes of war, industrial pollution, and conspicuous consumption. It was further weakened by the growing realization that resources were finite and serious external costs were inherent in the basic institutional structure of our economic system. Even the more complacent and optimistic members of society became aware of the "limits to growth" when the Arabs acted to remind them that petroleum resources, at least, are limited and growing more scarce.

Many public research efforts on energy will provide important findings for cooperatives. Much more systematic energy-use data will be available for planning purposes. Emphasis will be on energy use for processing and transportation where energy uses are high. Research to expand the use of new sources of energy will be explored.

Biological research will be heavily expanded, especially in the basic areas. Nearly every research priority evaluation has recommended expanding work on biological nitrogen fixation by plants and microorganisms in the soil; expanding genetic work; and improving photosynthesis where, even with corn, the efficiency of solar energy conversion is less than 5 percent.

Chemical research is dramatically changing notions about food. Ross Whitehead suggested that biological and chemical research could provide this meal for 1999: *"Appetizer,* fresh algae-fed oysters from a sea farm; *Entree,* 100% analog (spun soy-protein) chicken; *Vegetables,* mixed analogs. A casserole of tomatoes, cheese analogs, herbs, zucchini, and a base single-cell protein; *Dessert,* triticale cake, baked with flour from a hybrid of wheat and rye."

Research potential in agriculture staggers the mind. The following poignant example may exaggerate this potential, but it makes a point. Livestock diets may eventually include insect matter for protein. Fowl secrete an enzyme that dissolves the casings of insects. The key to turning insects into feed for other animals might come from application of similar enzymes during factory processing. The potential for insect-based feed is intriguing in part because fly pupae contain 61 percent protein and it has been calculated that a pair of houseflies starting in April could produce enough offspring by August to cover the earth 47 feet deep! This would be a lot of animal feed.[9]

Studies in health and nutrition will take on much more significance with current consumer and individual rights movements. Also, the worldwide problem focus of health and nutrition research will materially alter this type of research. It now appears that six out of ten leading causes of death in the United States may be degenerative diseases in one way or another related to nutrition. Nutrition research plans are broad. One recent inventory of problems to be studied by the USDA includes (1) What should people eat? (2) What are people actually eating? (3) How does what they eat affect their health? (4) What factors change people's eating habits? (5) What factors affect the safety, quality, and nutritional value of foods? (6) What can government intervention programs do to improve nutrition? (7) How do other programs and farm income policies affect nutrition? and (8) What are the responsibilities of the United States in improving nutrition in other countries[10]

Weather and climate research will focus on three areas of potentially profound United States impacts: (1) adaptation of crops, farming systems, water management, and soil use to adjust more effectively to variation of weather patterns;[11] (2) modification of weather with cloud-seeding and such techniques; and (3) significant trends in temperature, particularly in the latitudes of the upper United States and the possibly drastic shifts in production areas around the world.

Most cooperative managers agreed that scientific research and technological development were among the most important factors needed in their organization and in cooperatives generally. Academic people thought this even more strongly. However, the effective research and development were highly diverse and basically modest in size and program thrust. Data were somewhat difficult to standardize, but most organizations reported research and development expenditures at only about 1 to 2 percent of the total budget. Staff sizes ranged from essentially no designated scientific professionals up to as many as twenty-one, but in most cases the number tended to be in the two- to four-person level.

Little outside research work was hired. Also people identified as researchers were often more nearly quality control people.

Several of the organizations did collaborate in multi-cooperative research ventures. Only modest plans to increase research and development in the next decade were reported in most cooperatives. Managers, when interviewed orally, indicated somewhat more aggressive plans to increase the research program size than they did on their written response. Areas in which some modest projections were planned included product development, market analysis, and economic forecasting.

ASSESSMENT. Detailed case studies of the research programs were not made, but five plans, with some variation within each, evolved:

First, some organizations set up quality control programs, if needed, but did essentially no research and development work.

Second, others made sure that someone in the organization was knowledgeable in the research area but hired needed work done from outside private agencies.

Third, others used in-house staff members to do all needed research and development, except they were prepared to go to outside private agencies on highly complicated or specialized needs.

Fourth, some organizations maintained a small staff with reasonably good research knowledge who primarily monitored public research, especially at the land-grant universities and in the USDA; supported some small grant or graduate student funding at the public agencies; belonged to some of the multi-cooperative research groups; and interacted with other public and private researchers to learn about, or to borrow, as much as possible from them.

Fifth, others developed a high-quality, small research staff which kept a tight peerlike professional contact with public research groups including USDA utilization laboratories and land-grant university people, along with selected private research agency people; screened among needed research projects to assure that some of them were defined well enough to be picked up by public researchers; solved some feasible problems in-house; purchased some private consultation on projects needed but being done by public agencies; and kept a backlog of higher-risk projects for the public researchers, often encouraging the researcher with small grants of modest graduate student funding.

Research needed in expanding food production, marketing, development, and promotion tends to be expensive and promises to become more so.

Cooperatives have an opportunity to tap the public research community for many of their needs. Organizations using the fourth or fifth plan outlined above were faring quite well. Cooperatives can particularly look to other public agencies, as well as the traditional land-grant–USDA research organization, as more groups do food and nutrition research. USDA utilization laboratories seem to have several mutual interests with cooperative research staffs. The world network of research will open up new contact opportunities for some cooperatives.

Cooperatives depending on livestock products may need to do more research work. This would be done especially by concentrating on areas with broader cooperative interests such as the use of animal waste as feed for other animals, integrated contracting plans with patrons, and the use of additional marketing services for livestock farmers.

Cooperatives have a potential research problem in getting better prepared to promote new and evolving product lines necessary to assure specialized patrons an outlet for their product. Large conglomerate proprietary agribusinesses are reportedly reducing research and development outlays in their food lines. Cooperatives must consider multi-cooperative organizations such as the United Dairy Industry Association (UDIA) in the dairy industry.

Cooperatives must do long-term planning in research and development. Farmers will be influenced by long-term research developments. Scarcity in many supply inputs is a case in point.

Basic research needs typically are economically exploited with shorter and shorter time spans. Cooperatives must move quickly if they expect to compete in exploiting scientific breakthrough research.

Cooperatives will have more opportunities for joint ventures in the research and development areas and should explore multi-cooperative organizations and possibly joint ventures with public agencies. However, secrecy in many research efforts exists and will restrict this.

Cooperatives must look to the worldwide research efforts and their effect on the overall outlook for their particular organizations. Nutrition needs and nonfarm foods are being researched on a wide scale. Such research must be monitored by United States cooperatives. Cooperatives may need to do more of their own research on international trading arrangements.

Public Policy

Farmers are adjusting to their minority position in public policymaking, albeit slowly, and with apprehension and consternation. Their position has been eroding gradually since the beginning of the industrialization process, with a resultant decline in their numbers. They have lost political power. They must organize what political power they have in an effective and efficient way. Some farmers believe that the cooperatives must be used more effectively as the leading edge of the means to accomplish this. Others say this can be done by better coalitions between commodity cooperatives and the general farm organizations. Still others say that commodity cooperatives should not be involved themselves. Regardless, agricultural cooperatives need to be aware of their public policy obligations and opportunities.

DOMESTIC FARM PRICE AND INCOME POLICY. Farm policy since the 1920s has been concerned with chronic low returns to farmers. From the mid-1950s until 1972 food surpluses dominated the environment. Starting in 1972 many factors converged to give shortages that had profound worldwide repercussions, especially for that part of the world that had come to depend on United States surpluses. By 1977, however, farm conditions again returned to those of post-1972, and the farm income problem was reestablished as a surplus problem bringing great farmer unrest. There is controversy about the probable continuation of this surplus orientation, but most observers would advise cooperatives to expect it to continue.[12] Cooperative leaders also must know the more specific food marketing aspects of public policy. Agricultural marketing policy is under much current discussion. Cooperatives must expect such proposals to have implications of (1) changing size and specialization of farm units, (2) continued growth of sophisticated systems of agribusiness used to organize the procurement of farm products, (3) larger and more sophisticated and centralized cooperatives, and (4) farmer thrust for farmer security.[13]

V. James Rhodes summarizes the concern of the cooperative by this comment:

> The distinction between farmers' marketing and processors' procurement is more than a play on words. It reflects an important aspect of the complex issue of control in agriculture. While a dispersed open market model for farming has much appeal to farmers and agricultural economists, it does not provide a marketing system *for farmers*.[14]

Kenneth R. Farrell adds additional marketing policy issues of (1) the much closer ties of the entire world in food supplies and prices; (2) greatly changed retail food market institutions, particularly embodied in the "hypermarkets" which will merchandise food as a small part of the product line in a department store; (3) the low level of productivity improvement in the food industry, with resultant margin increases; and (4) the coming of age of the away-from-home food service industry.[15]

NATIONAL FOOD AND NUTRITION POLICY. Domestic price and income policy and food marketing policy will be the important parts of a total national food and nutrition policy for most cooperatives. But other broad policy matters should be well understood by cooperative leaders.

For much of the last decade a growing concern has evolved about the need for a national food and nutrition policy. This issue is now a major policy matter. Cooperatives may have a unique role to play in such policymaking. Several developments suggesting this are the increased marketing thrust of cooperatives, the concern of their members for important social issues, the possibility of increased working relationships between consumer cooperatives and agricultural cooperatives, and the need of a broader-based political constituency for policymaking in food and nutrition issues.[16]

Domestic food policy objectives have been identified as:

1) Efficiently produce an abundant, safe, and wholesome supply of food for U.S. consumers, 2) Provide for humanitarian needs at home and assist in meeting the needs abroad by sharing our abundance, 3) Maintain and increase the U.S. role in expanding world markets, 4) Maintain our dependability as a world food supplier, and 5) Increase farm incomes and improve rural communities.[17]

Consumer interests must be a part of any national food policy. The USDA Young Executives Committee in 1974 said:

A well-considered national food policy must reflect the long-run interest of consumers. Consumer needs are fundamental, and only by satisfying these needs can farmers maximize their returns in the long run.[18]

INTERNATIONAL FOOD POLICY. The world dimension of food policy will be important to cooperatives. United States commodity prices and resultant consumer food prices will depend heavily on the level of United States food exports and United States food aid policies. The importance of these issues is well known to most farm observers. They will not be observed in depth in this section, but cooperative concern with export policy is discussed elsewhere in this book.

One special problem of this increased world involvement affecting individual cooperative patrons and cooperative managers is price instability. Most observers consider this a major policy problem for this decade and one in which cooperative efforts can help farmers adjust. Positions on food reserves, trade arrangements, pooling systems, multi-cooperative organizations, and many others can be important in this issue.

Two other aspects of international policy impinge on the environment facing cooperatives. These are (1) the set of organizational and policy matters assured by the United States at the World Food Conference of 1974; and (2) the general long-run trends that are of particular concern in shaping world food policy.

The World Food Conference set up a coordinated set of world institutions to provide an active forum available for world food policy planning and negotiations. The rallying call of the conference was that "no child go to bed hungry or that no family fear for its next day's bread by the end of the decade." This has motivated a worldwide emphasis on food production that never existed before. The United States' role in this has been expanded.[19] Cooperatives may be asked to assume greater roles in carrying out these commitments, both in trade and in development.

The world food reserve issue is an illustration of a policy question brought about by such commitments which raises questions for cooperatives. An issue of development would be the active role of the United States government in fostering agricultural cooperatives throughout the world. This work is extensive and is being increasingly integrated into the worldwide economic development through cooperative efforts.[20] Many basic long-run worldwide developments affect our food policy. A dominant one is the fact that abject poverty persists for more of the world's people, and the corollary fact (relatively new) is that government leaders are attempting to do something about this through an interregional redistribution of world income and power. Also, the world economy has become more and more interrelated, but often among more sovereign and independent blocks. International cooperation is being strained, but new alliances are being tried. A decline has occurred in the economic dominance of the United States globally, with important implications for the United States food industry. This decline is accelerating.[21]

ANTITRUST. Antitrust policy issues are broad and pervasive in the food industry. The laws are general and unique in the degree to which they are subject to judicial interpretation and evaluation. This report provides only a limited view of this key environmental factor.

Some of the most relevant antitrust laws are the Sherman Act of 1890, the Federal Trade Commission Act of 1914, the Clayton Act of 1914 (amended by the Robinson-Patman Act of 1936), the Packers and Stockyards Act of 1921 (amended in 1958), and the Capper-Volstead Act of 1922.

A basic issue in antitrust is the monopoly power in the food industry. Proposals to deal with this are many. They range between two polar positions. One is a position arguing for a relaxed enforcement of the laws, and antitrust concentrating essentially on conduct violations (pricing violations, misleading advertising, etc.). The other is a position of selective restructuring of the food industry based on structural and performance rationale.[22] The Federal Trade Commission and Justice Department reflect a substantial movement to orient United States antitrust policy more toward structural, industry-type approaches and away from conduct, firm-type approaches. However, this may be swinging back some since 1970. Along with this approach the Justice Department, at least in its expressed concerns, has come down hard on cooperatives and their exemptions under the law.[23] The actual effect of antitrust is not always clear, but the effects on cooperatives specifically are worthy of concern for cooperative leadership. This concern has been stated by Ronald D. Knutson, Dale C. Dahl, and Jack H. Armstrong:

> Antitrust regulations applied to food industries are currently enforced in stringent—if not doctrinaire—fashion [in] cases of mergers or acquistions of firms at the same market level. Such enforcement has not stemmed the trend toward increased concentration, accomplished by internal firm expansion. At the same time, it has tended to force large growth-oriented firms to either expand vertically—backward or forward—or conglomerate into new product lines. Both types of growth have the potential not only to distort the structure of agriculture but also to lead to less competition, thus placing the producer in an even more disadvantaged position. In the long run, uncontrolled vertical integration has the potential of completely removing the independent farmer as a viable force in agriculture.[24]

The Justice Department's general antitrust enforcement policy has not normally been of great concern to cooperatives. Yet, cooperatives have been drawing Justice Department attention, whether warranted or not. Some such as Garoyan and Harris question the validity of this attention:

> It must be argued that if public policy remains supportive of the countervailing power concept of competition, efforts of federal regulatory agencies to curb the development of cooperative marketing associations would be counter to the public interest. Few if any farmer cooperatives are able to achieve profits comparable to large, general, food-manufacturing firms. Thus, it appears that current antitrust activity against farmer cooperatives is a deliberate choice of federal agencies, not because of performance, but because of ease and cost of filing complaints and court procedures against cooperatives, instead of tackling the large food manufacturers and distributors, including conglomerates.[25]

ASSESSMENT. Details of United States food policy are being hammered out. The parties to the compromise will be foreign policymakers, consumer interests, and agricultural interests. Price must be recognized as the rallying point for all these interests. Cooperatives must decide whether or not they will stand for their members on that issue. This position has been put forth succinctly:

> Consumers in the United States presently face no threat of food shortages as do their counterparts in many less developed nations. We have a wide array of options open, both to produce more food and to reallocate our abundant supplies of grain among alternative users if we need to feed more people. The central issue, with respect to food in the United States, is one of price, not availability.[26]

In such a policymaking forum, divisiveness within a minority is suicide. General farm programs will not become law if they disappoint and divide involved commodity groups. Neither will they become law if they fail to be compatible with other major programs such as food or energy, or if they make unreasonable demands in any single commodity or special interest provision. Concerted, informed group action by agriculture is mandatory for any substantive positive agricultural policy. Even this may be inadequate. If so, coalitions with other blocks of society will need to be sought.

Cooperatives must find their role in public policymaking. Don Paarlberg has defined four systems of policymaking—hallucination, confrontation, capitulation, and cooperation.[27] He rejects confrontation which causes power struggles and deadlocks from which retreat with honor is impossible. He chooses cooperation as the best. Harold R. Breimyer modified this and came up with a recommendation of a system based on compromise.[28] He claims compromise resembles cooperation and differs from confrontation in that each party must accept less than what is sought. Compromise requires that all parties accept a universal consensus of purpose and carry on lateral negotiations within it. Thus Breimyer collapses confrontation and capitulation into mutual capitulation, which is really compromise. Thus both arguments tend eventually to get to cooperation.

The evolution of a policymaking approach based on cooperation, given the nature of the policymaking environment, will in many ways mold cooperative survival strategies.

The next decade will provide farmers with alternatives among United States price and income programs. These will fall between a free market–export-dependent program without supply control and some type of program with greatly increased control. Control programs will probably fall among three options: (1) a monopolistic corporate agriculture; (2) a cooperatively oriented agriculture; and (3) a government-administered agriculture.[29]

Cooperatives must deal with commodity programs that fit into a total food and nutrition policy. United States price and income policy for nearly three decades has been an amalgamation of commodity programs superimposed on an undergirding of general price and income programs based either on supply control or a free market. How commodity programs would fare in a serious departure from such programs is not obvious.

Antitrust policy is becoming an increasing concern for cooperatives. The countervailing role of cooperatives in American competition and the special public sanctions accorded cooperatives are fundamental in antitrust policy. A most strategic issue ahead for cooperatives in the antitrust area is the amount of constraint that antitrust regulations will place on them directly. Such constraint will increase; how and how much will be hammered out. Therein lies a most strategic issue for cooperatives. Knutson, Dahl, and Armstrong have contended that existing regulations, properly enforced and administered, could go far in remedying problems of monopoly in agriculture. They see these opportunities mainly in improved enforcement of regulations currently administered by the secretary of agriculture, clarification and modification of antitrust policy in the food industry generally, and removal of government-authorized or imposed monopolies.[30] These authors also think that cooperatives can correct many of the antitrust problems themselves. This would mean a more prominent role for cooperatives as competitors and, in limited instances, they could become more effective bargaining agents on their own behalf. In the past, much of the aggressive marketing initiative for farmers has come from cooperatives; but the cooperative community is not changing enough to meet the needs of modern agriculture. The group approach increasingly has social constraint. The cooperative movement needs to be revitalized by strong leadership both inside the cooperative and alongside it.

Social Trends

The United States has a proud agrarian heritage. Changes in social environment test and at times cajole that heritage. Yet most Americans believe that agrarianism is important to our social structure. A cooperative builds on that social orientation and attempts to achieve a successful commercial enterprise. Cooperatives will find a thorough knowledge of their social environment of value in choosing the appropriate strategy for survival.

Actually the family farm that cooperatives protect is neither all commercial nor all social. Gulley has said, "The family farm today is . . . a commercial enterprise enmeshed in a highly coordinated national economy. The family farm has often been identified with freedom, independence, democracy, and progress, but these attributes have never been its exclusive property."[31]

Yet American agriculture, as Paarlberg has put it, "is being homogenized" into American society. The cooperative must recognize some loss in the visibility of its historically farm-production clientele. Two social issues—consumer movements and agricultural public relations—will be used to illustrate this change in a limited way.

In recent decades the consuming public has become more active in its concern about what, where, and how food is produced. This movement is commonly referred to as "consumerism." It has been brought about by four important factors:

> (1) Increasing consumer education and income, (2) product proliferation and complexity, (3) increased public awareness of the problems of environmental deterioration, and (4) changing social values relating to consumption and the quality of life.[32]

Consumers are demanding that the products they buy be fair priced, safe, and of good quality and are asking that they be given a voice in the decisions that affect consumers. Thus consumerism is forcing legislators, policymakers, and other government regulators to question the action of business in certain areas as well as the legal sanctions allowing such activity.

Business executives have had problems accepting consumerism. Webster concluded that business has come up with few planned and coordinated programs to respond to consumers. They would rather think that consumerism is something that does not affect them. Business leaders claim they have always been customer oriented and do not recognize consumerism as something much different. Thus the "do nothing" attitude of business in the area prevails or else business managers take an adversary position toward consumers.[33]

The response to consumerism by managers of agricultural cooperatives appears to be about the same as that stated above for business managers. Probably they were even more hostile and naive. In general, managers of cooperatives felt that consumer movements have not and should not have any influence on their decisions.

The depth of the consumer issue for cooperatives is questionable. Many authorities believe that it is underplayed and poorly related to by cooperatives in their appropriate social role. Yet Breimyer, discussing an array of crucial issues in the future environment of agriculture, said,

Although farmers are sensitive to consumers' attitudes, I put this topic well down on my list. One reason is that I believe consumer-farmer issues to be a side effect of other national issues such as inflation and foreign trade. In my opinion, consumers are less annoyed than apprehensive . . . the easiest way to reduce consumers' alarms will be to accumulate, once again, an appreciable reserve of feed stuffs and food.[34]

Consumers have placed nutrition high on this list of concerns. Public policymakers are responding to this social concern. Congressional initiatives have been many, but one of great interest to farmers is in the dietary standards area.[35] Several federal agency programs in nutrition have been affected by the recommendations of the National Academy of Science's *World Food and Nutrition Study* and other efforts encouraging the strengthening of both international and domestic programs of the United States government. Personnel and organizational changes of the USDA have been made to give greater emphasis to this area. The National Nutrition Consortium has long held for a broad national nutrition policy. President Carter's Food and Nutrition Reorganization study of the federal government recognized this social need as a factor affecting the organization of the federal agencies involved in food and nutrition.

Food assistance programs such as food stamps and school lunch programs are of consumer concern. During the last decade food assistance has been one of the fastest growing items in the federal budget. All of agriculture will need to be heard about these issues during the next decade. Cooperatives could lead in this discussion.

Possibly one of the most important consumer issues is the drive "to be heard." This has many forms but one of the most interesting was the appeal by the USDA's own Young Executives Committee which reminded the department that President Lincoln had termed it the "people's department." They concluded that "The Department of Agriculture has not adequately considered or involved the consuming public in its programs and decision-making process."[36] Moreover, they held that "well considered national food policy must reflect a long-run interest of consumers."[37] The increased changes by the USDA to answer these charges as started in 1977 are of much importance to cooperatives.

Environmental issues are of consumer and cooperative interest. Cooperatives must give increased attention to these. For example, topics on which cooperative personnel must be informed involve worker safety, foodborne illness, food additives, contaminants, pesticide residue, natural toxicants, and a host of others.[38] This highly controversial area is held by many to be one of the greatest threats to agriculture. Cost implications are many and are of concern to cooperatives.[39]

Public relations in agriculture have been widely discussed. Cooperative leaders gave this as an area where the farm community relates well to the social environment. But others are showing more concern. Concern in the area is of fairly recent origin but will continue during the next decade. Much of this is embodied in the issues raised in the consumer discussion above.

Yet some questions of public relations go to long-term conflicts that have existed among farmers, agencies serving agriculture, and within agricultural cooperatives themselves.

Great concern about public relations has been expressed, especially by the agricultural press. Despite this, both managers and academic people were nearly unanimous in saying that they thought the public image of agriculture in general was good.

The main problem voiced by managers was that the general public does not understand cooperatives. As far as contrasting cooperatives with corporations, managers believed "they [consumers] don't know there is a difference. . . . We [cooperatives] have never projected to the public the proper image for agricultural cooperatives."

The general feeling among managers was that cooperatives have to do more to educate the general public since "many people, especially the city people, have no idea of what a cooperative is or what it does."

Many of the programs visualized by the managers studied will test public support. The larger, more commercial thrust of cooperatives will not necessarily hold the support of the public. Several concerns about this issue are expressed in this book. The advertising programs of the National Council of Farmer Cooperatives illustrate cooperative concern on this problem, but not much other evidence was found that showed cooperatives responding to the problem.

ASSESSMENT. The cooperative is part commercial and part social. Cooperative strategy cannot ignore the changing social environment. The two social movements—consumer movements and agricultural public relations—are related and illustrate important issues of strategies for cooperatives, but these are only illustrations; other social movements can have major effects on cooperative strategies. Cooperative leadership must keep abreast of a wide array of changing social thoughts.

A basic consumer issue with specific implications for cooperatives involves a question: Can agriculture fulfill its needs to the world's hungry, and itself, and yet keep United States consumers reasonably happy with their share of the food and with their price for the food? The agripower philosophy of former Secretary of Agriculture Butz was clear in his decree to move foodstuffs abroad while favorably augmenting farm income and the United States balance of payments. John F. Kennedy put it more humanely in 1960, in a speech at Mitchell, South Dakota, when he said,

> I think the farmers can bring more credit, more lasting goodwill, more chance for freedom, more chance for peace, than almost any other group of Americans in the next ten years, if we recognize that food is strength, and food is peace, and food is freedom and food is a helping hand to people around the world whose goodwill and friendship we want.[40]

This issue continues to be of vital concern to cooperatives, and most authorities expect it to persist at least for the next several years.

Nutrition issues are high on the list of consumers' concerns. Cooperatives will feel the impact directly in this area as well as in many other areas such as natural resources, environment, and human health. Regulations in these areas are often enacted with minimum thought about cost. These costs, however, weigh heavily upon farmers and cooperatives. How the benefits to society are equated with the costs to society, and who bears those costs, should be of concern to cooperatives.

Cooperatives have inherited a problem of helping maintain a strong public image for agriculture. Many people interviewed thought that the image of agriculture and cooperatives in the eye of the general public was good. However, many other people are not so sure. Cooperative leaders have not kept the public aware of what cooperatives are and what they do. In the core sample, a majority of respondents agreed that there is public misconception about agricultural cooperatives. The question is, Why haven't cooperatives been able to educate the public? Most managers and academic people thought that agricultural cooperative leaders and others are not organized well enough to be effective in informing the public (both rural and urban) about agricultural cooperatives.

Managers did show some concern about the effect certain cooperatives have had on the public because of illegal political contributions. Also, concern was expressed about the evolving discussion that large cooperatives have used federal marketing orders to take advantage of consumers. Other than this, cooperative managers were not concerned, but academic people were quite undecided about the severity of the problems of the cooperative image.

Despite the lack of interest in agricultural public relations by the managers, all is not well with the way cooperatives are being treated and viewed. Cooperatives must look this issue in the face and view it correctly from the public's point of view. Any strategy that does not may be costly.

Economics

Cooperatives are economic organizations. Their success, growth, and general health are much influenced by the economic environment. Abrahamsen, in speaking of the factors affecting various periods of cooperative development, said,

> Current economic conditions, legal concepts, adjustments in agriculture, changing economic conditions, and aspects of worldwide social, economic and political forces influenced each of these periods.[41]

The prominence of economics in this notion of the process is obvious and impressive.

Economic history has always shown institutions to be products of their economic times. Those that adjust to new conditions survive; those that do not adjust die. New institutions evolve. The corporation, insurance company, supermarket, and cooperative all can be oriented to their time in history. Smart cooperative managers will innovate and adjust with strategies that fit the economic climate in which they find themselves.

The *raison d'être* of a cooperative is the economic and social welfare of the individual farmer. Too often a cooperative can lose sight of that. The managers interviewed tended to relate general economic conditions mainly to management functions of the cooperative as a business and did not relate farm policy or farm income situations to their strategies. They often failed to relate international economics to strategies and to recognize a possible role of cooperatives in using group economic relations to preserve the freedom of decisions by individual farmers.

They saw the farmer's survival mainly as his own business. Once he brought in his product or came in to buy supplies, they saw an obligation to act on his behalf, but that was mainly the emphasis. Management saw the cooperative more and more as a business and farmers either as its suppliers or its customers. Unless the economic situation affected the farmer simultaneously as it affected the cooperative as a business, it was not significant in the strategy process.

The economics of agriculture and food has a broad scope and is constantly changing. The scope is two dimensional. One dimension is the sheer size as measured by such factors as amount of land used, number of people and organizations involved, the part of the consumer's dollar spent, and the contribution to the nation's wealth. The other dimension is the great diversity of the system as measured by such factors as the many types of farmers and other organizations, the many services sold along with food, the significant place of food in world diplomacy, and the relationship of food to human health.

The economic scope is changing by simple growth and by a massive process of vertical integration. The effects of this changing scope permeate decisions of public policymakers, cooperative strategists, and individual farmers. Secretary of Agriculture Bob Bergland early in his administration described this for the public policymakers, saying, "Necessity demands that we have a comprehensive national food policy, not just a farm policy. This national food policy must address every step in the national food system, from farmers through to the consumer."[42]

The importance of this to policymakers is obvious, but its importance as a concept for cooperative leaders and the farmers lies in the fact that they also must view their strategy making in this broad perspective.[43]

RECENT ECONOMIC HISTORY. Agriculture was not always in the news economically but in recent years it has been.[44] Often people have tended to take food for granted. Agriculture has become a major growth industry, and issues such as world food demand, rising food prices, increasing agricultural exports, and health effects of food have contributed to its visibility.

Many people relate this change to conditions since the early 1970s, but the changes are from many sources—some new, and some operating for many years. From the late sixties into the early seventies, agriculture was a surplus industry. Farm incomes lagged in buying power. The economic situation was not critical, but certainly not exciting either.

During the end of the sixties, farm programs among other factors tended to balance production with demand. But many other forces were setting the stage for a new economic status of agriculture. These forces occurred in United States agriculture, in the United States general economy, and the world economy. The rate of United States agricultural productivity increase tended to level off. The general economy was strong and pulled excess resources out of agriculture. The excess capacity tended to disappear. By the late 1960s domestic food shortages and increased food prices were triggered by large areas of bad weather worldwide. Russia bought grain. Many United States regulatory and government initiatives clouded the picture.

During the early 1970s issues involving energy and other material shortages made for problems. Worldwide resource scarcity and environmental problems made news, dramatized by the Arab oil boycott.

The events of the late sixties and early seventies changed effectively the economic interdependence of agriculture and many other aspects of the United States and world societies. Following 1972, world population and income changes increased demand for United States agricultural exports.

Following 1972, the evolving forces of the late 1960s and early 1970s were compounded by new forces. Rapid growth in worldwide population and income, coupled with a worldwide commitment to attack world hunger, raised the demand for United States agricultural products. The planned economies added their demands. The United States food supply was not rising as rapidly as before, and the effect of research seemed to be leveling off. Major adjustments were made in exchange rates in world markets. Poor crops abroad added problems. The result was a prosperous United States agriculture from 1973 to 1975. Farm prosperity in 1973–75 raised food prices to consumers. United States export embargos confused world traders. The United States livestock industry collapsed. Land prices soared and debts in farmland were overextended.

By the end of 1977 the farm economy was about back to that of the pre-1972 era. United States farmers had harvested their third bumper crop and grain stocks were up. Production costs went up rapidly. Farm debt skyrocketed. Grain prices declined, and the government was back into the farm subsidy business. Farm income dropped sharply. Farmers were restless and got much publicity from widespread strikes. The agricultural situation improved in 1978. But after a decade of highly variable, complicated, and often disappointing economic conditions, the food industry had concern about the future.

A Congressional Budget Office study found four important effects of these times on farmers:[45]

1. Farmers faced greater dependence on export markets. The study held that "the long-term prospect for continued U.S. grain exports at these levels depended on factors largely beyond the control of the U.S. government—worldwide weather conditions and policies of foreign government."

2. Income gains were unevenly distributed. Livestock producers were

hurt badly and larger farms gained more than others. Production costs and land values increased. The study held that these factors were probably now built into farmer expectations.

3. Food prices contributed to overall inflation. The study said that from 1971 to 1974, 40 percent was attributable to food price increases.[46] This image could well haunt the food industry, especially farmers, for some time.

4. The structure of agriculture changed drastically. The study contended that many factors had contributed to fewer and larger farms, but the relative stability and reduced uncertainty fostered by government commodity policies and the favorable treatment of larger farmers were important.

This economic environment has involved United States farmers, imposing great stresses on them in production and management. They have been thrust into a position of prominence as a fundamental industry. They have not done badly in continuing to raise their standard of living. Their numbers have been reduced and many are big business managers. They survived it, but they answer the bell for the next decade like a boxer who leads on points, but who was knocked down three times in the last round, and has three more rounds to go.

The Future

PRICE AND INCOME. The general economic health of the country has great bearing on the level of prosperity of cooperatives and their members. Historically, this relationship has been a basic assumption underlying agricultural price and income predictions. This may be altered some in the future. Schuh in his provocative article on the new macroeconomics of agriculture challenges the rigid and ironclad assumptions many make about the historical relationship between the general economy and the agricultural economy.[47]

Schuh holds that the agricultural outlook must now be viewed in an open world economy, not in the closed economy assumed in most past analyses. He contends that the closed perspective "tends to ignore linkages to an international economy and focuses instead on the secular income problem that agriculture faces as it is subjected to the process of economic growth."

He concludes that

> this perspective no longer seems relevant to the major problems now facing U.S. agriculture. Outmigration from agriculture has declined markedly, long-sustained growth in factor productivity appears to be leveling out, and major shocks to the U.S. food and agricultural sector have come from abroad as a result of the American economy having become more open. At the same time, the rules governing the trade and exchange relations among countries have changed, and there is pressure for even further change as the less developed countries push their demand for a new international economic order. . . . Looking ahead, the major analytical and policy problems agricultural economists face

will have to do with how U.S. agriculture and the economy as a whole fit into a rapidly changing, interdependent world economy.

He further adds that the recent recession

> was probably the first time in our modern history that a severe recession in the general economy had little effect on agriculture and was due in no small part to a strong export demand as well as the continued broadening of our welfare programs.

Among his several conclusions is the rather optimistic one that "the secular income problem in agriculture is now largely behind us."

Neither Schuh nor other analysts will argue that the agricultural economy and the general economy are unrelated. The new argument is about the type of relationship and the implications of that relationship.

Most predictions are for a reduced rate of growth in the gross national product (GNP) with a rate of 2–4 percent compared with an average of about 4 percent from 1960 to 1973.[48] Per capita real income is expected to increase at slightly less than the GNP rate. The labor force increase of recent years will drop from about 2–3 percent to 1 percent by 1985.

The inflation rate is expected to be 4–8 percent annually. This compares with higher rates recently but from the end of World War II until 1973 the rate was 1–3 percent.

President Carter in his Economic Report of the President transmitted to Congress in January 1978 said, "Our economy is basically healthy." He set out four goals for the economy over the next four years:

1. "We must continue more steadily toward a high-employment economy. . . ." He hoped for a GNP growth of 4½ to 5 percent per year with a reduction of unemployment of one-half of 1 percentage point each year.

2. "We shall rely principally on the private sector to lead economic expansion. . . ." Five out of every six new jobs were expected to be from the private sector.

3. "We must contain and reduce the rate of inflation. . . ." The president was more qualified on expectations here, but hoped to reduce inflation below the stubborn rate of 6 to 6½ percent existing after the high rates of the early 1970s.

4. "We must act in ways that contribute to the health of the world economy." The chronic rate of inflation in the industrialized countries, which is much above that of the United States, and the weakness of the United States dollar are trouble spots in this area. On balance, most experts would say that the general economic outlook is favorable, if not buoyant, for the next few years.

AGRICULTURE. The domestic demand for farm products is expected to increase at a rate of 1.0–1.3 percent, or only slightly more than one-half the rate of increase during the last decade. Population increases will be down from recent periods and will be less than 1 percent per year. This will be offset some by a slight upgrading in diets.

The foreign demand will be increasingly important and more difficult to predict. The effect of foreign demand on United States cooperatives and their members will lie in economic policies regarding food aid, the potential of commercial exports, and the rate of agricultural development of other countries, especially the lower income ones.

Food aid programs will continue and may increase some. But the main impact of foreign markets will continue to be in commercial exports, not increased food aid under subsidized programs. The food aid program will follow the ebb and flow of world politics and will still be a strengthening factor for United States producers. Cooperative leaders should stay in touch with this side of the market, but overall it will have about the same effect in the next several years as in the last several.

Farm exports are expected to grow erratically at a rate of 3–4 percent annually. This rate of increase will not be easy to maintain. Cooperatives will have to change to help hold that rate. About 25 percent of total farm receipts will come from exports even though trade policies will probably be more restrictive. Trade liberalization has been pushed hard, and additional freeing up of trade will take strong negotiation. Agricultural interests, including cooperative leaders, will need to be in the middle of these negotiations if they expect agriculture to fare well.

Exports will continue to be erratic due to variation in weather, economic conditions, purchase patterns of the socialistically planned economies, and the fits and starts of competition from the lower income countries.

World agricultural development is an important subject and increasingly will affect United States farmers and their cooperatives.[49] We are dependent on these countries as markets for United States goods and services; the lower income countries have been buying about 35 percent of all United States exports and 40 percent of the agricultural exports.[50] They take 50 percent of our wheat exports, 60 percent of our cotton exports, and 70 percent of our rice exports. In the mid-1970s the lower income countries provided over 25 percent of United States raw material imports and over 50 percent of our food imports. Sales of United States exports to these countries have been increasing more rapidly than have those to the richer countries.[51] Several studies have shown that in the countries in which we have contributed substantial technical assistance to help them with agricultural development, our exports to them have gone up sharply. On the other hand these countries are developing and they want to develop trade outlets. They will become strong competitors for United States farmers in some products.

Despite the fact that United States agricultural exports have increased at rates unthinkable prior to the 1970s, much apprehension exists about the future. Food needs in the world are great and will increase drastically. Yet, at least 80 percent of the world's population increase will come in the poor countries. Their ability to pay will be a question. The thrust will be to try to help them to help themselves. Also sales to USSR and Red China are politically sensitive. The threat of war, the increased variability in world markets, and lack of knowledge about these new world markets are all causes for concern. Also, developed country policy, especially the Common Agricultural Policy of the European Community, shows only modest liberalization toward accepting United States imports. The United States could continue to find itself a residual supplier, on the whip end of many worldwide political and economic developments. United States dominance as a food reserve base only partially softens these effects. Foreign demand is important and growing, but the rate of growth of agricultural exports is not at all a sure thing.

On balance, the rate of increase for total farm product demand will be from 1.3 to 1.6 percent per year which is somewhat lower than the rate of recent years. Total world food demand will be up more than that and could increase the United States demand, depending on how much of that market increase the United States gets. United States farm output continues at a high rate, but our output per unit of input has been slipping. The United States agricultural plant can produce much more, possibly as much as 15-20 percent more over the next decade if economic conditions foster it. The increase was 2 percent per year from 1940 to 1973 even with 50-60 million acres idled by farm programs. World output potential is high, but the problems of getting that production are formidable.

The real income situation for United States farmers under these conditions can improve some, but the pressure will stay on. Variability in prices and incomes will be greater, and the resultant effects of that fact on a highly capitalized agriculture will be more severe. The assessment of the future will be more difficult and more complicated, especially by international conditions. Inflation will exist, but it will not favor agriculture as much as it has historically. Farm programs will play prominently in the future of price and income situations. The profit position of farmers in different regions of the country will vary more. The profit position will vary more among individual farmers, with the richer ones becoming increasingly richer relative to the poorer ones.

Nonfarm sources of income have been much more important to farmers and this trend will continue, but possibly at a declining rate. During the late 1960s, farmers started getting more than one-half of their total income from nonfarm sources; this averaged 54 percent of the 1966-70 period and increased to nearly 59 percent by the end of the 1970s.[52]

Although the per capita disposable income of farm households has long been below that of nonfarm households, this difference has been narrowing and actually converged temporarily in 1973. The improvement that has occurred in this situation should be maintained, but farmers will have a difficult time gaining relatively more compared to nonfarm people.

GENERAL AGRICULTURAL ECONOMIC TRENDS. Many general agricultural trends affect cooperatives. This section will discuss four.

First, the relationship between rural and urban societies is a major issue. It affects land and water use, farm policy, community responsibility of local cooperatives, sales policy of cooperatives especially to nonmembers, along with a host of other concerns of cooperatives. This interrelationship becomes more important to local cooperatives as they deal with day-to-day local living, to lawmakers as they deal with resource use and other problems nationally, and to policy experts as they try to decide how to use our natural resources to help our neighbors internationally.

Secretary of Agriculture Bergland summarized the concern of national growth by saying,

> Balanced national growth . . . implies two things: . . . a strategy . . . avoiding the self-defeating trap of approaching urban and rural problems and development totally independent of one another. . . . A strategy of parallel growth and development not only to meet economic needs of people but equal emphasis on meeting the social and cultural needs as well.[53]

Second, the structure of agriculture will continue toward fewer farm people, fewer but larger farms, and probably more nonfarm rural neighbors.

The number of farm people went from about 15 million in 1960, or 8.7 percent of the total population, to just about one-half that many farm people in 1977. And those 7.8 million were about 3.6 percent of the total. This will continue, but at a slackening rate.

The number of farms continues to drop, but at a declining rate. This will continue. Total farmland continues to decline some but not enough to offset the drop in farms, so size of farm continues to grow. The average United States farm at the end of the 1970s had about 100 acres more than the average one had going into the 1960s.

One of the most dramatic changes has been the drop in the proportion of farms operated by tenants. In 1935 tenant-operated farms peaked at 42 percent but dropped to 11 percent by 1974. During that period the percentage of farms operated by full owners rose from 47 percent to 63 percent and those by part owners from 10 to 26 percent. Part owners are the smallest group but are the bulwark of commercial agriculture and control most of the resources used in farming.

The new structure of agriculture is of vital importance to cooperative planning. In 1976, 6 percent of the farms had farm sales of $100,000 or more, and they accounted for 60 percent of the cash receipts from farming and made 40 percent of the realized net income. On the other hand, the farms selling under $10,000 accounted for 61 percent of total farms, had only 5 percent of the receipts, and accounted for about 16 percent of the net income. The following data illustrate these facts for these two groups and other sizes of farms:

Farm Sales Classes	Cash Receipts	Realized Net Income	No. of Farms
	(Percent of total)		
$100,000 and over	60	40	6
40,000 to 99,999	20	23	11
20,000 to 39,999	10	14	11
10,000 to 19,999	5	7	11
Under $10,000	5	16	61
All farms	100	100	100

Source: *Farm Income Statistics,* July 1977, USDA.

These general economic trends with regard to farm people and the makeup of farmers are of long standing, and their pressure will continue. New data on the trends are always readily available and should be used by cooperative leaders for planning purposes. United States agriculture will be composed of much larger and more complicated farms. The nature of this change will be hammered out in the economic environment predicted. Two well-known authorities illustrate the alternatives. Breimyer put it this way:

I have been saying that the 1970s will be the decade of debate of who will control, and the 1980s the decade of final decision. I see no reason to change that calendar. Unless it becomes national policy to do something about it, by 1990 commercial agriculture will be in large hands, the majority of them nonfarm. Only small part-time farmers will show staying power.[54]

Paarlberg put it this way:

The family farmer will slowly give up his historic role of supplying all production inputs. . . . Agricultural production will require farms so large . . . that a single person will be unlikely to supply them all.

In commercial agriculture, the nearest thing to the family farmer will be a farm operator who lives on the land with his family, rents his farm, borrows his money, and hires his labor. He will make his own decisions on combining these inputs and contracting for both inputs and output.

Besides those farms that produce most of our crops and livestock, there will be part-time farms, combining the production of food and off-farms with rural living.[55]

Academic people interviewed were asked to evaluate the impact of agricultural structural changes on cooperative strategies. Some of their answers shed light on this question:

Must continue to grow—do more tasks then they have in order to succeed in the future.

The family farm will get bigger and more specialized. The co-ops can adjust to the first easier than the second. There will continue to be a million or more part-time, moonlighting farmers, who will market a small proportion of many products and a large proportion of a few such as beef calves. These farmers are nearly indestructible in a competitive sense. They present special problems to co-ops but perhaps some special opportunities as well. There will be further encroachment by large corporations—both in direct production and by contractual integration. These threaten cooperatives.

They will need to evaluate their services to win support of large farmers.

Co-ops must be more business management conscious. Services and products must be priced in relation to cost of servicing customers of various sizes.

We will soon have 250,000 farms and ranchers producing ¾ of total agricultural production. We may need to have some local co-ops specifically designed to serve the large producers—others to serve small producers. The diseconomy of trying to serve both types from one co-op may be excessive. The regional cooperatives will be able to serve both types of locals.

I believe it will strengthen them. This large farmer will demand more but I also believe he will fully understand the importance of his cooperative and therefore will make a total commitment to it. This probably will strengthen the large *well run* cooperatives.

The increasing proportion of larger farms could be a factor in curtailing cooperative growth. The larger farms are in better position to market or buy on their own. This could be a factor in curtailing the cooperative volume of business and efficient corporations. Cooperatives need to adjust to serve these large farm firms.

Third, American agriculture becomes increasingly resource intensive. This is especially important with regard to uses of energy, land, water, and capital. Much discussion will prevail about energy use in agriculture. Actually the proportion of the nation's energy used in farm production is quite low; it is much higher in food processing and distribution, and even higher in food preparation in the home or at other eating establishments.

Yet researchers and policymakers will be dealing with energy use in agriculture, and cooperative planners should follow these developments. Many think that pressure on our relatively limited land resources will increase as we move toward the end of this century. Water availability is already a severe problem in many areas.

The year 1973 dramatized the growing scarcity of resources worldwide and the effects which that can have on world food production. Historically, our agrarian heritage and our orientation to domestic food needs resulted in policies assuring available resources for our food production. The food system will need to fight more and more aggressively for its resource needs. This is not just a domestic issue; so long as the United States is the main source of food for many other countries, especially in periods of shortfalls, we must have resources available for food production.

Capital shortage is an issue. And it is broader than cooperatives or agriculture. In the decade of 1955–64, the American economy utilized $760 billion in capital; in 1965–74, the consumption doubled to $1.6 trillion. A composite of estimates indicates a need for the huge sum of $4.5 trillion ($400 billion of this for agriculture) during the 10-year period (1975–84). This means the nation's total supply of capital must rise at a compounded annual rate of 8.7 percent during the next decade compared with 6.7 percent in the last decade.[56]

Estimates of investment requirements are not far apart, but projections of savings vary widely.[57] Government budgets can add to the level of savings if tax collections outweigh expenditures. This potential surplus is the principal uncertainty in the forecasts. A small government surplus (combined federal and state budgets) is predicted in most studies. Thus most predict no shortfall in overall savings relative to investment needs. The Brookings Institution's Duesenberry-Bosworth study concludes that a capital shortage can be avoided in the late 1970s, "but just barely." The study assumed no major shifts in government policy; however, higher inflation would combine with real GNP growth to boost government revenues by 11.7 percent per year, resulting in a net budget surplus.

Wallich concludes that a capital shortage can be avoided, at least over the next few years; but the assumption of a surplus, or near surplus, in federal and state government budgets is crucial to that conclusion.

Capital needs of the United States economy are staggering, and agricultural capital needs are of great concern to cooperatives. Also lenders are concerned about the growing risk of farm loans; a recent USDA study showed that more than one-half the lenders think the risk has increased.[58]

New technology, inflation, and the desire of farmers to increase their incomes have greatly increased farmers' capital requirements. Average production assets per farm between 1959 and 1975 have been growing at a compounded rate of 9.2 percent. Production expenses and gross farm income per farm increased 9.1 and 8.6 precent respectively.

Fourth, food has become a broad dimension in our life-style—sort of a symbol of social concern. The youth of the 1960s condemned many parts of the food system, including our food eating habits, as antisocial worldwide. Many of these notions still exist—for example, the movement toward vegetarian diets. Concern with food to some degree has been a symbol of consumer emphasis and could be a focal point for more consumer movement emphasis. Affluence has made food a symbol of service. We now eat more than one-third of our meals out. Food is a symbol of what's new. The product line of the supermarket is expected to change daily. Nutrition research is tending to make food a symbol of health.

ASSESSMENT. This assessment will discuss ten general developments. Most economic trends will have similar effects on farmers and cooperative management. Some will not; where that is true, it will be pointed out.

First, the economic environment of the next several years will continue to increase the common interest between cooperatives and their patrons. This may sound incongruous since the tendency of cooperative people has been to equate the two interests. But outside observers have made telling arguments that cooperatives have often been entities in and of themselves and farmers' interests were not so obviously up front. This discussion does not assess the seriousness of that leverage, but it does argue that several trends should make this problem less of a problem in the future.

Greater relative investment by farmers in their cooperatives will add to the common interest. The food industry has moved sharply toward a nonfarm food industry. Several studies have shown that at least 80 percent of the food system's contributions to the GNP is from nonfarm activities. This among other economic facts will move farmers to a broader and more integrated organization so they can tie into the huge complicated overall food system. The investment will increase in absolute terms and probably relative to the farmers' total assets.

Farmers' needs increasingly must be paramount in cooperative planning. The process of holding on to some cooperative public sanctions will mean more scrutiny on this point. It is a common rally point among the increasingly diverse group of patrons in most cooperatives. Farmers express greater concern about economic activity and policy than was observed in discussions with the managers. Managers will need to become more interested in this issue, and this will increase mutually between farmers and the cooperatives.

Cooperatives may be a more obvious factor in aiding farmers in the battle of who will control agriculture. Some experts argue that farmer cooperatives are the best institutions to use in public policy for this purpose. This would bring the cooperative organization and its patrons closer together.

Both management and boards will need to increase their leadership role in the type of environment predicted above. However, this makes it imperative that patron support be strengthened. The survival of cooperatives will depend on greater appreciation of this mutual interest between cooperative and patron.

Many of the processes followed by cooperative management are more aggressive, more risk-taking, and more nonfarm oriented than the rhetoric that sometimes pervades producer relations meetings. Patrons must be brought into the process with open and aggressive producer meetings. Strategies must be of mutual interest and mutually carried out.

Second, economics increasingly has become more important to cooperatives. This was widely held by managers. This trend will be accelerated. The scope of issues discussed above makes this point. The impacts of many of these issues are discussed elsewhere in this book. However, inflation is one trend that must be emphasized.

Inflation is predicted as a fact of life, and cooperative strategy choices and operational tactics must deal with it. Not only must policy choices continue to recognize the effects of a given rate of inflation, but they must recognize the fact that cumulative effects of such high rates of inflation create a host of new management problems.

High consumer prices and inflation can bring adverse public criticisms of the cooperative or any business organization. This can compound public relations problems.

High rates of inflation and resultant resale prices make for more discrimination on the part of buyers and can result in quick and substantial shifts in consumer acceptance of individual items. Cooperatives must add this variable to their strategy about product and service issues. Also, this will test patron loyalty.

Farmers historically have expected inflation to aid them. Their prices have gone up first, and this gave them a competitive edge. This situation may not exist in the economic environment predicted. Both management and patrons must be more cautious about this variable in the planning process.

Transportation and related costs are singled out as being vulnerable to inflation. Cooperatives may want to emphasize this variable in planning their procurement and distribution policies. This may affect their evaluation of a strategy including agreements among cooperatives.

Third, a well-understood and forthright strategy will be needed if cooperatives are to survive in the economic environment ahead. Different cooperatives will need to face this future in different ways. Cooperatives as a group have some common concern in this environment. Joint planning is possible by cooperatives, and this is a special advantage they have. They must search for effective mutual strategies. The two cooperative needs—that of the individual and that of the group—may not always be compatible. Much time will be spent trying to find strategies to fit these needs as the environment changes.

Fourth, integration is an economic fact of life and its importance will grow in the future. Integration is one of the overpowering concepts of this half century. Many innovations have occurred in how to apply integration; many more will evolve in the future. The cooperatives studied were using vertical integration toward consumers, particularly into processing. Cooperatives have had early experience in integration forward from the farm into the marketing functions. Noncooperative businesses with which cooperatives increasingly will compete have had a wider variety of experiences with integration and are more effectively integrated in the more advanced stages of the food system than are cooperatives. Cooperatives must have experts in integration if they expect to make integration one of their most used strategies.

Fifth, this economic environment calls for better techniques and tools for cooperative leaders if they are to make the type of analysis necessary for survival. The level of economic analysis observed was not impressive. The emphasis too often was with rather traditional price and income prediction models, which gave results that the managers themselves said were not important to the cooperatives' future. The longer-run, more integrative type of economic planning being employed by noncooperative management was lacking. Two recent USDA studies in prediction can be useful to managers since they were undertaken more to illustrate what should go into an analysis than to make a specific prediction.[59] They provide a useful broadening concept that would improve most models observed in the cooperatives studied.

Economic environment is a broad topic and backdropped much of the interview time with cooperative managers. Yet the specific economic analysis in cooperatives was nominal. Managers reported their concern about better data. Detailed questioning was done on the type and frequency of economic analysis by their staffs. In general, almost all rigorous analysis was short-run of one year or less. Two organizations claimed to do longer-run, in-house, general economic analysis. Others claimed to monitor that done by outside analysis for short-term forecasting and evaluation.

Managers were asked to predict the future. Most were basically optimistic about the economic outlook for the next decade not only for farmers but for cooperatives in general and their own organizations specifically. Interestingly, management in only one organization showed any real concern about facing an unfavorable economic outlook in the next decade.

The managers said the farm economic climate was a factor in retention of membership and required some specific planning, particularly for possible shortages of inputs or other emergency type issues. However, they believed that the general business outlook had similar impacts on management of cooperatives and of noncooperative businesses. This was more relevant than the effects of the economic environment on the farm operations themselves.

Managers predicted inflation rates generally in line with professional economic forecasts; most saw an annual rate of 5–7 percent over the next decade. Most were quite optimistic about farm incomes and agricultural export levels over the next decade.

Sixth, general public policy will have a greater impact on cooperatives in the future than in the past. Such policy increasingly will have an important economic component. This follows a long evolution of this trend, effectively presented in studies by Wayne D. Rasmussen of the USDA.[60] J. B. Penn and others have illustrated the current application of this point.[61] The international dimension will be equally heavily laden with economic policy.[62]

Seventh, the international dimension of cooperative responsibility requires an understanding of the large and complicated international economic environment. Many cooperative leaders have had no obvious reason to become experts in this area. The importance of the international environment to cooperatives specifically will increase. More important, the increasingly important effect of the international economic environment on this country generally means that anyone wanting to understand either agriculture or economics must understand that part of the environment. The shifting cooperative advantage of certain foreign producers will have important impacts on given groups of United States producers. This can also affect cooperative policy, especially where diversified patron membership exists.

Eighth, cooperatives will be subjected to increased economic variability. This will come in many ways and will affect patrons and cooperative managers individually and together. The cooperatives can figure heavily in providing a buttress against economic shocks and actual shortages of supplies for patrons.

Cooperatives may need to get more into reserve supplies as United States and world food reserve policies are hammered out. Regional variation in the profitability of agriculture within the United States will increase. The recent plight of low incomes for farmers varied widely by regions. The fact that food has become a broader dimension in our life-style brings new variability.

Economic conditions are varying more widely, and food is subjected to a wider array of more complicated factors. These two facts make it inevitable that the economics of food will give farmers and cooperatives increasingly complicated problems. Shortages, such as dramatized by energy, figure heavily in most United States economic forecasts. Since cooperatives are owned by their farmer members, they have a unique concern and opportunity in taking care of their regular patrons in both the short- and long-run planning horizons. Shortages and volatile prices can provide cooperatives with special opportunities. Many progressive cooperatives have used this to their advantage, especially in the inputs area over the last three or four years. Cooperatives must not relax; planning should retain that advantage. Merely hoping for fortuitous circumstances to return could be disastrous.

Ninth, cooperatives will be affected by changing rural-urban relationships. The stresses and strains of competing demands for resources can mean a new mediating role, especially between rural nonfarm residents and farm residents. This can be compounded for supply cooperatives who sell to a large nonmember business. Also, the rising nonfarm sources of farmer income can have merchandising and financial implications for cooperative management. Special social needs such as those of low-income farmers might call for new cooperative approaches both domestically and abroad. The relationship between farm and consumer cooperatives could be affected by the growing interdependence of farm and nonfarm relationships.

Tenth, technological development and economics always have been interwoven. This relationship will continue. As farmers strive to adapt to new research and constraints such as energy availability, their cooperatives may need to do more to service them in these areas. Also, the need for technological transfer and adaptation abroad will give new opportunities for cooperatives.

2 Competitive Environment

TYPICALLY, cooperatives neither dominate their competitors nor are they dominated by them. The key role of cooperatives has always been to make agricultural markets more orderly and competitive. At best, they have tempered competition rather than dominated it. Current developments raise the question as to whether or not such a competitive posture is adequate for cooperatives to survive.

Today's patron is expecting more from the cooperative. He expects a diversified and coordinated bundle of products and services. He expects technical and economic efficiency whether reflected in timeliness of service, price of product or service, modernization of processing and distribution facilities, or economy and efficiency in a host of other areas.

Many cooperatives are considering new ventures, especially vertical integration into marketing. The competitive environment for such ventures is vastly more complicated and demanding than the environment formerly faced by cooperatives. Such ventures intensify use of capital, specialized expertise, and purchase of agency or customized services.

Cooperatives figure heavily in the agricultural policy question of who will control agriculture. One alternative, especially put forth to preserve the independent farm, is a strengthened cooperative system. Such a focus would put the cooperative even more directly into the spotlight of total competition between agriculture and other parts of the economy.

Cooperatives are innovating with new institutions. A range of new institutions has been built on a broad base of local cooperatives and centralized regional cooperatives. This includes such institutional innovations as federated regionals, multi-cooperative (interregional) commercial ventures, multi-cooperative service or political ventures, international alliances, and joint ventures.

These developments raise questions about the competitive intent of cooperatives—especially in the processing of farm products and the manufacturing of farm supplies and their competitive interaction with others. Also, these developments bring public scrutiny of cooperatives such as is usually reserved for noncooperative companies.

The above trends mean larger and more complicated cooperatives. Such cooperatives will push harder against the competitive environment. Cooperatives can expect the competitive environment to push back. More public scrutiny is assured.

The economic and social environment will impinge on cooperatives in much the same way as it will on any other agribusiness. Cooperatives will have some special concerns for the unique interests of their patrons, such as maintaining a market outlet or providing a service. They will have a special role of farmer interest in marketing operations such as price making, price reporting, public policy program administration, and implementation of environmental and health regulations.[1] Increased foreign trade of agricultural products and inputs extend cooperative competition internationally.

Farm Suppliers

The farm supply industry has become of increasing importance in the United States food and fiber system. In 1950, 42 percent of the materials farmers used were purchased; by 1972, this had increased to 62 percent.[2] In 1976, farmers spent nearly $82 billion for production ingredients, distributed as shown in Table 2.1.

The supply industry is large and concentrated (Table 2.2). Agricultural supply firms form an important but not a dominant segment of it. Cooperatives are increasing their activities in some supply areas such as petroleum and fertilizer.

This industry is so broad and complicated that only a brief, almost cursory, statistical treatment of the competitive situation is possible for the following four segments of it.

PETROLEUM. Fuel and oil input is a small but important proportion of the total purchased production inputs. Fuel needs have increased for machines,

TABLE 2.1. Distribution of farm production expenses, U.S., 1978

Type of input	Percent of total expenses
Feed and seed	19.9
Repairs and operation of capital items	10.2
Fertilizer and lime	7.6
Livestock purchased	7.1
Hired labor	8.5
Depreciation	17.0
Taxes	4.3
Interest on farm mortgage debt	4.9
Rent	5.0
Miscellaneous	15.5
Total	100.0

Source: USDA, *Agricultural Statistics, 1977,* U.S. Government Printing Office, Washington, D.C., 1977, p. 471, preliminary.

TABLE 2.2.   Industry concentration ratios for selected agricultural inputs

Classification title	SIC code number	Value of industry shipments accounted for by the four largest companies						Factors adjusted for†
		1970	1966	1963	1958	1947	1966*	
Fertilizers	2871	30	34	34	34	36	(50-60)	R,B
Fertilizers, mixing only	2872	24	21	20	18	18	(40-50)	R,B
Agr. chemicals	2879	47	36	33	. . .	. . .	36	. . .
Petroleum refining	2911	33	32	34	32	37	(65)	R,B
Lubricating oils and greases	2992	39	36	36	29	36	(50)	R,B
Farm machinery	3522	40	45	43	. . .	. . .	(70)	B
Industrial trucks and tractors	3537	52	49	54	52	57	(60)	B
Food products machinery	3551	21	21	22	16	18	(30-40)	B
Tractors	3521	. . .	. . .	. . .	69	67	(71)	. . .

Source: Alex F. McCalla and Harold O. Carter, "Alternative Agricultural and Food Policy Direc-
tions for the U.S.: With Emphasis on a Market-Oriented Approach," Preliminary report
prepared for Policy Research Workshop on Public Agriculture and Food, Price and Income
Policy Research, Washington, D.C., Jan. 15-16, 1976, p. 12.
* Adjusted concentration ratios.
† R—markets are mainly regional in scope.
B—census industry definitions are too broad.

grain drying, and fertilizer production. Estimated agricultural use of total refined fuels was 5 percent in 1963.[3] In 1969, total farm use of liquid fuels and LP gas was estimated to be 7.6 percent of total United States use.[4]

The petroleum industry can be characterized as concentrated and fragmented, geographically. In the mid-sixties, 85 percent of the oil and 90 percent of the gas brought to the surface were produced in seven states.[5]

The industry has a high degree of vertical integration, with the major segments consisting of crude oil production, refining, marketing, and transportation. In 1963, 32 of the largest oil companies produced 65 percent of the total 2.8 billion barrels produced in the United States, and 151 companies operated 304 refineries. By 1966, the number of refineries had declined to 286, and 48.6 percent of the refineries were owned by the 23 largest companies.[6]

In 1970, about one-half of the crude oil production was produced by about 20 firms. In refining, the 4 leading firms accounted for nearly one-third of the total value of shipments. In some regions, adjusted concentration ratios based on value of shipments for the 4 largest companies were estimated as high as 65 percent.[7]

The cooperative share of the petroleum market (based on percent of farm supply expenditures) has increased from 19 percent in 1950-51 to 31 percent in 1974-75. Cooperatives operate 8 refineries with a crude oil distillation capacity of about 400,000 barrels per day or three-fourths of their needs. But this is only around 3 percent of the United States total. In 1975, cooperative-owned oil wells furnished only 10-12 percent of the crude oil refined by cooperatives.[8]

FERTILIZER. Commercial agriculture uses about 90 percent of the total United States fertilizer tonnage.[9] For the fiscal year ending June 30, 1976,

36

TABLE 2.3. Annual production capacities for specified fertilizers or derivatives and concentration ratios of the top four firms, top eight firms, and cooperatives with operating plants, January 1, 1976

Fertilizer or derivative	Total capacity	Top four firms	Top eight firms	Cooperatives
	1,000 tons	*percent of total capacity*		
Anhydrous ammonia	18,567	24	40	20
Urea	5,887	36	56	28
Nitric acid	11,958	37	54	16
Ammonium nitrate	8,692	26	46	17
Ammonium phosphate	5,132	42	62	28
Phosphate rock	61,100	50	73	0
Phosphoric acid	9,089	38	65	25
Potash	11,699*	51*	79*	3†

Source: Duane A. Paul, Richard L. Kilmer, Marilyn A. Altobello; and David N. Harrington, *The Changing U.S. Fertilizer Industry,* Agr. Econ. Report 378, ERS, USDA, Washington, D.C., Aug. 1977.
* Includes the United States and Canada.
† U.S. capacity.

fertilizer consumption in the United States and Puerto Rico was 49.1 million tons. Almost 21 million tons were of primary nutrient content (nitrogen, N; phosphate, P_2O_5; and potash, K_2O).[10] Use in 1976 of anhydrous ammonia, superphosphate grade, and potash materials was 4.9, 1.4, and 4.0 million tons respectively.[11]

United States firms listed as having the capacity to produce synthetic anhydrous ammonia increased from 7 in 1940 to 71 in 1969 and then decreased to 59 in 1976. In 1966, the 4 largest firms had nearly 26 percent of the total production capacity. Twelve firms or almost one-fifth of the firms had more than one-half (53 percent) of the total capacity for anhydrous ammonia production.[12] On January 1, 1976, the 4 and 8 largest firms (with operating plants) accounted for 24 and 42 percent of total production capacity respectively (Table 2.3). Two cooperatives were in the top 4 or top 8 and accounted for 12 percent of U.S. total annual production capacity.[13]

Fifty plants owned by 45 firms were capable of producing ammonium phosphate in 1966. By January 1, 1976, only 28 firms with 40 plants were producing ammonium phosphate. The top 4 and 8 firms accounted for 42 and 62 percent of the total annual production capacity, respectively. The 4 largest firms included 1 cooperative with 12.5 percent of annual production capacity.

Potash production capacity was held by 10 firms in 1965. Forty-eight percent of total capacity was in the control of the 3 largest firms.[14] By 1976, the 4 largest firms accounted for 64 percent of total annual production capacity.

In 1966, cooperatives had about 14 percent of the total domestic

A local farm supply cooperative in Navarre, Kansas.

COOP FEED SERVICE CENTER

capacity of anhydrous ammonia. By 1976, the cooperative share of United States anhydrous ammonia production capacity had reached about 20 percent. That same year cooperatives were producing nearly 25 percent of the United States wet-process phosphoric acid—a 21 percent increase in production since 1959-60. In 1974-75, it is estimated that cooperatives sold to farmers 30 percent of the total fertilizer and lime purchased.

FEED. In 1975, 5,590 United States feed manufacturing establishments were producing at a level of 1,000 tons or more each of formula feed.[15] By region, the Corn Belt—Ohio, Indiana, Illinois, Iowa, and Missouri—was the largest producer of formula feed, accounting for nearly 18.5 percent of the total formula feed produced in 1975.[16]

In the feed manufacturing industry during 1964, the 4 largest firms accounted for 17 percent of the total sales; the top 20, 31 percent. By 1969, industry sales of the top 4 slipped to 12 percent of the total and the 20 largest had slightly under 22 percent.[17]

In 1969, cooperatives operated 1,810 feed manufacturing establishments producing 1,000 tons or more of formula feed with total production of more than 20.7 million tons. By 1975, cooperative establishments had declined to 1,412 with production of 20.4 million tons. In 1969 and 1975, cooperatives accounted for 20.6 and 19.6 percent of the total production respectively (Appendix Table A.5).[18]

PESTICIDES. Farmers consume approximately 40 percent of the pesticides marketed in the United States. Value of pesticides for farm use was estimated at $519 million in 1965,[19] and $759 million in 1970.[20]

Production of basic pesticide chemicals is carried out by some of the largest and most diversified firms in the country. In 1964, 169 plants produced basic pesticide chemicals, and 1,542 plants were involved in formulation and distribution of pesticides. About 50 percent of the total pesticides (in terms of sales) was produced by the top 4 companies in 1964.[21] By 1970, 33 firms reportedly accounted for about 80 percent of the total.[22]

Cooperatives have not been in the primary production of pesticides until recently when 1 association acquired a plant. However, cooperatives do own and operate around 30 formulating plants. In 1971-72, 3,433 cooperatives handling pesticides accounted for 20 percent of the farmer's total purchases of pesticides. During 1974-75, cooperatives handled 30 percent of farmers' total purchases of pesticides—a 10 percent increase in 4 years.[23]

MACHINERY AND EQUIPMENT. United States agriculture has become more mechanized. The value of farm machines and equipment shipped for use in the United States was nearly $7.0 billion in 1975—a 265 percent increase over 1964. Almost 42 percent of the value of farm machines and equipment in 1975 was accounted for by farm tractors; harvesting machinery was second with 17.7 percent of the total.[24]

Concentration in the farm machinery industry is quite high. Seven major manufacturers ("full-line" companies) operate on a worldwide basis. These companies are diversified, and farm machinery accounts for about 50 percent of their sales.[25]

In 1963, the 4 and 8 largest firms accounted for 43 and 55 percent of the value of farm machinery shipments respectively. By 1972, about 50 percent of the world farm machinery sales was accounted for by the 4 largest firms, and 61 percent was accounted for by the 8 largest firms.[26] United States tractor sales of the top 4 contributed 83 percent of the total.[27]

The role of farmer cooperatives in the sale of machinery and equipment to farmers is small. In 1974–75, cooperatives handled about 2 percent of farmers' total expenditures for machinery and equipment.[28]

In a 1970 study of the extent cooperatives franchise and distribute farm machinery in the United States, it was found that only 105 farmer cooperatives were franchise dealers for major farm machinery manufacturers.[29]

ASSESSMENT. *The key competitive issue for cooperatives is whether they can become strong enough to compete with the large firms that operate in the farm supply industry.* The supply industry is large and concentrated. Stages involved are raw material procurement, manufacturing, blending, storing, wholesaling, and retailing. Concentration is rather high at the procurement and manufacturing stages; it is high but not as important in the other three stages. The industry is diverse in products and services, and this mix changes continually. Consequently, many of the noncooperative supply firms, often conglomerated and handling other products and services, impact on the strategies of cooperatives. Some strong multi-cooperative organizations have penetrated this area of big business, but generally cooperative activity has not yet reached a scale comparable with non-cooperative establishments.

Competitive cooperative strength could come from many different means. Vertical integration back into raw material discovery and procurement is one. Another would be horizontal growth in selected specialized input sales areas. Cooperatives can diversify in many ways within a range of input items as well as by coordination of marketing, supply, political, or other service divisions. Markets can be expanded to new customers within both the farm and nonfarm markets. Also, geographic market expansion is possible. Multi-cooperative or joint venture with noncooperatives might strengthen cooperative programs.

The means for increasing competitive strength exist. Cooperatives must decide whether to use a stronger stragegy.

A related issue is how far should a cooperative diversify to compete in the input industry. What types of organizations, functions, and products go together? Establishment of long-run strategies for cooperatives involves an appropriate relationship between supply and bargaining and/or processing and marketing cooperatives. Farmers must decide whether their cooperative can effectively handle two or possibly three very different competitive fronts. Procurement of supply in the inputs industry is well established and strong, often with international operations. The retail side is diverse in products and services. In many markets it is rather highly concentrated. Does the cooperative have an inherent competitive advantage because its owners are also its users? How effective can an agricultural cooperative be in a complex and different competitive structure?

Sheer size issues are important. Farmers must decide if they want their cooperatives to put together such a large program as is necessary to compete in this environment. Does the cooperative want to be this big? Does it want to consolidate institutions, often designed to perform specific functions, into a large conglomeration where their purpose and identities will be much less recognizable and accountable? To compete in this type of supply industry at any one stage—raw material procurement, manufacturing, blending, and storing, wholesaling or retailing—requires that the strategy be carefully formulated. To compete at all levels, and then to undertake the marketing of member products, are functions that were probably not anticipated by most cooperative founders and framers of cooperative legislation. A cooperative board may also need to consider the possibility of establishing cooperative control over farm production for processing needs or increased marketing programs to protect markets. When all these issues are put together, they multiply the number of crucial questions for agricultural leadership.

Special backup problems for cooperatives for these decisions include acquiring the necessary capital, acquiring the necessary research and development, maintaining quality control, gaining consumer entree, managing needed inventories, and carrying out appropriate distribution.

Cooperatives must also assess the international issues of the farm supply industry. Should they go into selling inputs abroad? Growth opportunities exist there. Also, should they expand importing crude oil and other supplies or even explore further the possibilities of producing crude oil in foreign countries to help ensure a source of supply for their American farmer members? Many noncooperative firms are assessing this market and making long-range plans accordingly. Procurement opportunities for supplies have broad international implications. Farm machinery has been internationally widespread, but the great need for intermediate technologies by foreign farmers may not be met by this industry. Foreign agricultural cooperatives are strong in many countries. Cooperatives are an acceptable form of business enterprise in many areas of the world whereas noncooperative organizations may not be as acceptable. Many service needs of agricultural people abroad are not being met, and no obvious institution is evolving to supply these needs. Some cooperatives' leaders interviewed were looking at these alternatives.

Many cooperative arrangements have allied a general farm organization (a political arm) with the first-sale, commodity marketing arm. Many more recent developments have aligned service cooperatives, such as insurance, travel agencies, and financial organizations, with the general farm organizations. Supply cooperatives were more conventionally oriented toward farm input goods. A key planning question is, What does the competitive environment say about consolidation of all such supply and political organizations in a state or region?

An issue arises about the kind of multi-cooperative or other joint arrangements cooperatives should build in the inputs industry. Some important successes have evolved, and general satisfaction and enthusiasm were expressed by many cooperative managers for the multi-cooperative ventures in which they were members. However, the rapid growth of several ventures was starting to be of concern. Many issues such as the allocation of scarce capital among multi-cooperative organizations and other uses are yet to be resolved.

Opportunities for joint noncooperative and cooperative ventures must be studied. Possibly these can share the large cost, help spread the power concentration, and facilitate comparative advantages for each party. The type of institutional arrangements needed here, whether all-cooperative or joint with noncooperatives, are large, expensive, and quite different from the arrangements under which cooperatives have usually operated in the past.

Supply cooperatives must decide who their customers are to be. The requirements of different farmers for inputs and related services are a problem. Also, cooperatives appear to have a varying and somewhat unconvincing strategy with regard to servicing nonfarm customers. Service needs of these customers can vary. The bigger farm patron has different service and product line needs than the smaller one, and the urban patron may have needs different from either of the farm patrons.

Cooperatives in the supply industry have problems within the structure of their own membership. Examples exist of a few large patrons pulling out of the traditional cooperative to make their own joint purchases. The question of whether the cooperative obtains purchasing commitments from its own membership will figure heavily in any strategy to enhance the supply cooperative power base. The diversity of customers for supply items raises questions of patron equity and loyalty. Also, a strong power move to non-farmer markets by supply cooperatives could raise public questions as to whether this was the reason for farmer cooperatives.

Supply cooperative strategy may need to deal with the relationship between consumer cooperatives and supply cooperatives. Many similarities exist. Changing United States social structure could foster consumer cooperatives, and consumer cooperatives are already important in many parts of the world. More attention needs to be given to possible strategies of coordinating the activities of consumer and supply cooperatives.

The issue of available raw material is a crucial one for cooperatives. To have access to raw or processed supply is often crucial, but it raises a major issue of how much should cooperatives vertically integrate backward into raw supplies or processing capacity.

Cooperatives have moved significantly into raw supply procurement pressured during scarcity and fluctuating prices. Whether cooperative integration or coordination of farm inputs with processing or marketing of raw farm products is advantageous raises complex organizational questions. How far to integrate backward to tie farm production to farm supplies is an issue. Some have contended, for example, that cooperatives should have been much more active in integrating the broiler industry through their feed supply base before the noncooperative firms moved so far.

The resources needed to compete in the discovery and procurement of raw material—for example, in fertilizer or oil—are massive. The opportunity cost for capital used here versus the value of assured supplies is a major question. This is especially difficult while an overall United States energy policy is evolving.

The role of public treatment in the farm supply industry may be an issue. Cooperatives have fewer public sanctions in this field than they do in other areas. Most public treatment for farmers is focused on the farm product bargaining, processing, and marketing areas and not on the input and procurement areas. Supply cooperative managers must think twice about inferring special strengths of public support from marketing experiences, especially concerning integration and multi-cooperative organizations.

Public scrutiny recently has raised important questions in related industries such as oil and tires. This has not yet significantly affected cooperatives, but it could become an issue. Cooperatives can perform some countervailing power role in this field at local levels, but they must not overestimate their potential effect nationally. A question arises as to whether it is strategic for farmers through their cooperatives to make large investments in an industry depending primarily on the public to control the balance of competitive power nationally.

Antitrust policymakers may see reasons to encourage cooperatives in the input industries to foster competition. The role the farm supply division holds in a large noncooperative conglomerate is often minor. This minor role makes the more specialized supply cooperative competing with this unit vulnerable. Problems of the smaller specialized company competing against a larger diversified one are well established. When the supply division of a conglomerate firm is also relatively small within its organization, the conglomerate management may force conduct on the supply division that imperils the competing cooperative. This array of problems could call for some evening up of market power, especially by strengthened supply cooperatives.

Another cooperative issue is whether to exploit the growth momentum of the supply industry. The farm supply industry has been a growth industry and promises to continue to be one for some time. New organizational arrangements often find growth industries much more favorable than stable or declining ones. On the other hand, a growth industry can compound problems for a related industry and individual institutions in it. Also, an existing institution in the growing industry can have growth-related problems. Cooperatives must distinguish where they are with regard to these farm-supply growth factors.

United States farm production has been energy intensive and thus input-supply intensive. This could be altered some, but few are arguing that a major shift will occur. The increasing supply needs of farmers may give cooperatives some opportunity to preempt noncooperative firms from entering the newer expanding supply areas. Should such a strategy be exercised? Should cooperatives serve needs other than those of their farmer members? These are key cooperative questions.

Strategic planning will involve the question of greatest comparative advantage for an individual cooperative. Would the growth advantages of the supply industry be more promising as an alternative to an individual cooperative than pushing into the marketing side of the food industry where growth may be more sure, but probably more variable? Demand for some major farm products may be much more unstable than demand for the farm supplies and complementary supply markets. On the other hand, cooperative leaders feel much more responsibility for monitoring markets for farm products than they have either for preserving availability or costs of supplies or for creating and fostering demand for supplies. A question exists as to which will mean the most to patronage income and farmer needs.

Growth of supply cooperatives into wider geographic markets may be an alternative for some. Cooperative input markets have been fairly well restricted to historical membership areas. This is not a criterion used by noncooperative competitors. Most cooperatives studied seemed to project a continuation of this policy, but some were expanding. The plans of the expansion-minded cooperatives worried the others.

Some good reasons exist for expansion of farm product markets outside the patron areas; to do so with farm supplies will be more difficult. Yet some cooperatives studied were comparing expansion opportunities geographically with expansion opportunities within existing marketing areas. Some were using both. Growth in local and regional markets may be open to cooperatives. Cooperatives need to do careful local market analysis. But they should use care in designing strategies based primarily on regional or national structure data. The share of market by most cooperatives is important in local markets. Data were not collected on market shares, but shares are known to be high enough in some markets that possible social restraints must be kept in mind when seeking more intensive sales in a given market. Overall, cooperatives appeared to have growth potential left both locally and in expanded markets.

Unsuccessful ventures by large noncooperative establishments into the farm supply industries have resulted in sales of some outlets to cooperatives, and some managers saw promise that large supply conglomerates might yield more business in the rural specialized market. Cooperatives may well have a competitive advantage because they have special entree to their patrons who are also their owners. Cooperatives may find that rural supply distribution opportunity is also useful in any program by cooperative structures to control agriculture. Cooperatives' strength and their growth potential in supplies argue that any cooperative system designed to control agriculture will substantially involve the supply cooperatives.

Overall, the competitive environment for the farm supply cooperative is complicated and often rather severe. But cooperatives have proved their

44

capability in this area. They are established in many products and markets. They are innovating and growing. Many cooperative plans involve expensive and long-term commitments. Farmers must continue to study this environment and determine carefully when their cooperatives can best fit into it.

First Handlers

Cooperatives, from the beginning, have been active in the first sale of farmer products. Farmers historically sold their own products. They somewhat reluctantly gave up that function. Cooperatives have taken on many aspects of marketing for farmers, and individual farmers now depend heavily on experts hired by cooperatives to do much of their marketing for them. Tradition in cooperative marketing relies heavily on this first handling—a function not a part of farm production but obviously close to the farmer's heart. Some believe government sanctioning of cooperatives was intended only to bridge this gap. Such a view could imply, at the least, that other marketing functions should be sanctioned on a case-by-case basis. Regardless of view, the cornerstone of cooperative strategies is in the first-handler function.

The first-handler competitive situation is a crucial one in assessing equity for farmers. It has been a basic target for those evaluating marketing cost, price formation, institutional innovations, and competition in general.

Cooperatives have had to grow to handle the increased farm output. Farm output, from 1960 to 1976, increased 22 percent, or 1.4 percent annually. Crop and livestock production were both increasing until 1972. However, increased demand for grains, especially stimulated by exports, resulted in higher prices beginning in 1972. Livestock production then began to decline, but crop production continued fluctuating upward.[30] In 1976 estimated value of cash receipts for United States farmers from farm marketings for total livestock and products was $46.4 billion and for total crops $47.9 billion.[31] These receipts came from a variety of commodities and called for diversified marketing services by cooperatives (Table 2.4).

The kinds of markets used by farmers have changed over time. Around 1900, livestock sold off the farm were assembled by local dealers and shipped to terminal markets. Packers would then bid for the livestock at terminal markets. Since the early 1900s, however, the importance of the terminal markets has been declining.

In 1950 more than 25 percent of cattle, 60 percent of hogs, and nearly 43 percent of sheep and lambs were purchased by packers directly from the farm, auctions, or country dealers. By 1975 these percentages had increased to 66, 72, and 74 percent for cattle, hogs, and sheep and lambs respectively.[32]

Markets for feed and food crops at the first-handler level have remained relatively competitive. Most off-farm sales of these crops are handled by a large number of country elevators.

TABLE 2.4. **Distribution of cash receipts from total livestock, livestock products, and crops marketed, U.S., 1976**

Commodity or commodity groups	Total cash receipts from livestock products	Total cash receipts from crops	Total cash receipts from livestock, livestock products, and crops
	percent		
Cattle and calves	41.9		20.6
Hogs	15.9		7.8
Sheep and lambs	.8		.4
Dairy products	24.6		12.1
Eggs	6.8		3.3
Broiler and farm chickens	6.6		3.3
Turkeys and other poultry	2.1		1.0
Wool	.2		.1
Other	1.1		.5
All livestock and livestock products	100.0		49.1
Food grains		14.2	7.2
Feed crops		28.1	14.3
Cotton, lint and seed		7.4	3.8
Oil-bearing crops		18.5	9.4
Tobacco		4.7	2.4
Fruits and tree nuts		7.3	3.7
Vegetables		11.0	5.6
Other		8.8	4.5
All crops		100.0	50.9
All livestock, livestock products, and crops			100.0

Source: USDA, *Agricultural Statistics, 1977,* U.S. Government Printing Office, Washington, D.C., 1977, p. 468.

In calendar year 1974, for example, country elevators purchased 75.4 percent of the corn, 79.8 percent of the wheat, and 80.2 percent of the soybeans sold by United States farmers.[33]

In mid-1975 country elevators operated nearly three-fifths of United States commercial storage capacity. The estimated total storage capacity of various United States commercial operations was as follows:[34]

Type of Operation	Share of Total
Country elevators	59.4
Subterminal elevators	6.6
Terminal elevators	17.0
Export elevators	2.7
Processors	9.0
Feedlots and poultry producers	5.3
Total	100.0

Milk shipped from the farm has moved greater and greater distances. The system had become much more integrated. Several functions such as country receiving and cooling are no longer necessary. To some extent, first-handler functions in milk are only a means to get the product into a form or place for more profitable functions. Cooperatives are important receivers of milk, but integrated functions have mostly replaced the older first-handler functions.

Farmers through their cooperatives have been increasing their share of products marketed. Most cooperative activity is at the first-handler level. Cooperatives, with few exceptions such as milk, still do not market a large share of farm products (Table 2.5). Total share has been increasing and additional functions of processing and marketing are being added. By individual commodities, however, cooperative share at the first-handler level of livestock and poultry has declined.

Cooperative strategists must recognize that farmer interest in first-sale and cooperative orientation historically has been on this function. Strategies that continue to emphasize this first-handler orientation, possibly by strengthening it through integration, may be more prudent than strategies that emphasize functions farther toward the consumer.

The two points of power at the first-handler level are the specialized nonintegrated marketing cooperative and the large vertically integrated food processors. Several cooperatives have moved away from this historic, competitive battleground. Yet many cooperatives, especially first-handler ones, are still rather specialized in one commodity. This group includes those handling dairy, cotton, livestock, and grain products. The dairy cooperatives have increasingly integrated into processing and distribution, and some livestock and cotton ones have moved this way. With cooperatives holding much of the first-handler business and with large noncooperative food processing companies integrating backward into assembly, the assembly step in food marketing has become rather concentrated at the local level in many markets.[35]

In 1970 an estimated 22 percent of total United States farm output was produced under production contracts and vertical integration. This compared with 19 percent in 1960.[36] The use of production contracts in 1970 was greatest for sugarbeets, processing vegetables, and seed crops (Table 2.6).

TABLE 2.5. **Farm level share of products handled by farmer cooperatives, by selected commodities, 1950–51, 1960–61, 1964–65, 1969–70, 1975–76**

Commodities or commodity group	Share of Market				
	1950–51	1960–61	1964–65	1969–70	1975–76*
	percent				
Cotton and cotton products	12	22	25	26	26
Dairy products	53	61	65	73	77
Fruits and vegetables	20	21	25	27	25
Grain and soybeans	29	38	40	32	44
Livestock and livestock products	16	14	13	11	10
Poultry products	7	10	9	9	7
Other	15	22	25	27	24†
Total	20	23	25	26	29†

Source: Gene Ingalsbe, "What's the Cooperative Market Share?" *Farmer Cooperatives,* FCS, USDA, Washington, D.C., Feb. 1977, p. 4.
* Taken from Lyden O'Day, *Growth of Cooperatives in Seven Industries,* Cooperative Research Report 1, ESCS, USDA, Washington, D.C., July 1978.
† USDA unpublished data, preliminary.

TABLE 2.6.   Percent of U.S. farm output produced under production contracts and under vertical integration, 1960 and 1970

Commodities or commodity group	Production contracts		Vertical integration	
	1960	1970	1960	1970
	percent			
Crops:				
Feed grains	.1	.1	.4	.5
Hay and forage	.3	.3	...	...
Food grains	1.0	2.0	.3	.5
Vegetables for fresh market	20.0	21.0	25.0	30.0
Vegetables for processing	67.0	85.0	8.0	10.0
Dry beans and peans	35.0	1.0	1.0	1.0
Potatoes	40.0	45.0	30.0	25.0
Citrus fruits	60.0	55.0	20.0	30.0
Other fruits and nuts	20.0	20.0	15.0	20.0
Sugarbeets	98.0	98.0	2.0	2.0
Sugarcane	40.0	40.0	60.0	60.0
Other sugar crops	5.0	5.0	2.0	2.0
Cotton	5.0	11.0	3.0	1.0
Tobacco	2.0	2.0	2.0	2.0
Oil-bearing crops	1.0	1.0	.4	.5
Seed crops	80.0	80.0	.3	.5
Miscellaneous crops	5.0	5.0	1.0	1.0
Total crops*	8.6	9.5	4.3	4.8
Livestock and Products:				
Fed cattle	10.0	18.0	3.0	4.0
Sheep and lambs	2.0	7.0	2.0	3.0
Hogs	.7	1.0	.7	1.0
Fluid-grade milk	95.0	95.0	3.0	3.0
Manufacturing-grade milk	25.0	25.0	2.0	1.0
Eggs	5.0	20.0	10.0	20.0
Broilers	93.0	90.0	5.0	7.0
Turkeys	30.0	42.0	4.0	12.0
Miscellaneous	3.0	3.0	1.0	1.0
Total livestock items*	27.2	31.4	3.2	4.8

Source: Ronald L. Mighell and William S. Hoffnagle, *Contract Production and Vertical Integration in Farming, 1960 and 1970,* ERS-479, ERS, USDA, Washington, D.C., Apr. 1972, p. 4.
* The estimates for individual items are based on the informed judgments of a number of production and marketing specialists in the U.S. Department of Agriculture. The totals were obtained by weighting the individual items by the relative weights used in computing the ERS index of total farm output.

Vertical integration was greatest in sugarcane production—60 percent of the crop.

In the area of livestock and livestock products, in 1970, 95 and 90 percent of the fluid-grade milk and broilers respectively were produced under production contracts. Vertical integration was highest in eggs—20 percent of total egg production (Table 2.6).

ASSESSMENT. *Will the competitive market endure?* is the big question for the first-handler cooperative. First-handler cooperatives were formed to provide a competitive outlet. They have relied on an open competitive market. However, with increased integration by food conglomerates and some cooperatives, accurate and equitable price discovery and reporting are threatened.

To function effectively, a modern market agency must monitor these markets for accurate, needed market information. Both cooperative and noncooperative managers can be sure that the organization with the best information will have an advantage.

Farmers may need to depend heavily on accurate market information and public sanctions such as federal orders to assure orderly marketing. In some commodities such as livestock and grain, farmers depend on large central markets to perform orderly marketing. These markets have been supervised to some extent by public scrutiny. Can cooperatives continue to expect the public scrutiny necessary to maintain open markets? Can they expect the public sanction necessary from federal orders and other arrangements to maintain orderly administered markets? Whether farmers can continue to rely heavily on public administration to maintain orderly markets will influence choice of general strategy.

Noncooperatives may not compete directly with the cooperative for first-handler business for some products or geographic markets. The large noncooperative is interested primarily in dependable raw materials and consumer accessibility over the longer run. If the conglomerate can be assured of farm products at appropriate times, it may prefer to leave the assembly function to cooperatives.

Cooperatives must select the organizational strategy that can buffer the power base of the large conglomerated food organization. The cooperative may well need the competitive buffer of the fringe noncooperative companies, especially the independent ones. However, it is not clear where the real competitive power lies. The first-handler cooperatives must decide whether to worry about retailer power, conglomerate processing power, differentiated suppliers with national brands, or the diversified and integrated cooperatives. Any one of these may be the real competitive force of concern to the first-handler cooperative. Important competitive issues will continue to prevail at the local level. Integration may be necessary for a cooperative to maintain the backup strength to bargain at that level. It is questionable whether the specialized nonintegrated marketing cooperative can compete.

How specialized should the first-handler cooperative be? Such cooperatives can add products or services that allow more services and more options for the disposal of products handled—for example, providing information, hedging expertise, financing, and advertising. Also, many have considered adding more products to their assembly and handling organizations. Some are contracting with producers for products produced or for breeding stock.

Increased foreign trade has raised questions about the small, specialized cooperatives. The large proportion of grain assembled by cooperatives, compared to their small percent directly exported, is well known.[37] Torgerson has said, "An aggressive search will be made to develop a combination of marketing mechanisms and policy tools that will provide for more orderly marketing in the international arena *and assurances for returns* that will allow producers to expand their productive efforts."[38] The competitive structure at various levels of product handling is a formidable issue for the predominantly first-handler cooperative.

The specialized cooperative at the assembly level competes with a large diversified conglomerate in many instances. The specialized cooperative faces special problems. This situation at the assembly level is analogous to the well-known situation favoring the diversified operation over the specialized one at the processing or distribution level.

The first-handler cooperative has often been tied to the political arm of a state or regional general farm organization. As political issues become more national in scope and farm policy more complicated, strategy planning for the first handler becomes more difficult.

Is efficiency enough? is much more than a trite question for the first-handler cooperatives. Such cooperatives must typically rely mainly on serving specialized needs and achieving top efficiency.

Whether the size and scale of efficiency for the first-handler cooperatives can be expanded through use of new marketing methods such as the teleauction raises many questions. Does this expand the size and scale of efficiency for the first-handler cooperative? Can a cooperative innovate more effectively with some of these new techniques than can a large bureaucratic conglomerate? The specialized cooperative must look carefully for such significant strategic advantages with special services and returns to its patrons.

The first-handler cooperative must not exist as just another outlet; it must fill a specific need. Many cooperatives have been started to provide an outlet when a competitive outlet ceased to exist. Yet many studies show excessive numbers of grain and livestock firms (from an efficiency point of view) operating in individual markets. The plight of too many, too small local cooperatives is well known. The competitive structure now evolving at the first-handler level will demand increased scale of operation.

First-handler cooperative activities are close to its patron owners. To control much of the action, cooperatives will probably need to keep a reasonable proportion of the business at this level. It is not clear whether competition will intensify at this function. Large noncooperatives may think that they must have this function to guarantee their supply. If this turns out to be true, competition for the specialized local-handler cooperative will be severe. Even if first-handler cooperatives have first call on this assembly function, it is not obvious that it will be relatively profitable. Most studies show that the assembly function by itself yields low returns. Often, it is a necessary evil for accomplishing other profitable functions.

It may be necessary to add additional commodities or additional services to make a successful cooperative. If this does not bring about a viable economic unit, it may be necessary to integrate into more of the marketing functions. Once integration is started, the dilemma arises as to how far to integrate. Thus the competitive structure at the first-handler level is difficult to assess and more difficult to predict. It will be changing. Cooperative strategies here are crucial and carry much beyond the concept of local cooperative efficiency. Most strategies must be based on local conditions, but first-handler cooperatives face new challenges. Their strategies must be subject to rigorous, periodic review.

Food Processors

Food processing has grown from a rather specialized, craft-type enterprise to a large-scale, capital-intensive enterprise of strength and complexity. Many food processors were founded on specialized products, processes, and patents. Today they are highly integrated and conglomerated. Many companies in the field are multi-national. Cooperatives have not been heavily engaged in processing, but this study found great interest and important cooperative planning of new ventures in this area.

Food processing is not an easy function to describe. W. Smith Greig gives one of the more encompassing definitions: ". . . a large, diverse, increasingly integrated, growth industry, with decreasing numbers of firms and establishments, not particularly concentrated (in terms of other industrial segments of United States industry) but increasing in concentration (but not particularly in 4-firm or 8-firm ratios), with increasing diversification and conglomeration. In fact, a substantial portion of the food processing industry consists of some of the world's largest multi-national conglomerate-industrial firms."[39]

In fiscal 1964–65, 19,895 corporations that were primarily food processors reported to the Internal Revenue Service. In fiscal 1970–71, 16,285 corporations reported—a decrease of more than 3,600 in five years.[40] The decline in number of companies has occurred in almost all segments with few exceptions—such as beet sugar (Table 2.7).

In 1975 value of shipments for food and kindred products was $172.5 billion—16.6 percent of total value of shipments ($1,041 billion) accounted for by all manufacturing. The food industry in value of shipments is the largest manufacturing industry. The transportation equipment industry ranked second with nearly $113 billion in value of shipments.[41]

Within the food and kindred products, meat and dairy products accounted for 25 and 13 percent of total value of shipments respectively. In meat-packing, the number of companies declined from 2,833 in 1963 to 2,293 in 1972. Concentration in the largest 4 firms and largest 8 firms declined, but the 20 and 50 largest firms accounted for 51 and 66 percent of the value of shipments from meat-packing plants respectively—an increase of 2 and 6 percent since 1963.[42]

In the fluid milk industry, the number of companies has declined almost by half—from 4,030 in 1963 to 2,025 in 1972. The value of shipments, however, has increased from $7 billion in 1963 to $9.4 billion in 1972. The largest 4 companies' percentage of value of shipments has declined from 23 to 18 and the largest 8 companies' percentage has declined from 30 to 26. The 50 largest companies nevertheless accounted for 56 percent of the value of fluid milk shipments in 1972—an 8 percent increase over 1963.[43]

In 50 selected food and beverage industries (Table 2.8), more than 50 percent of the 4 largest firms increased the percentage of their respective industry's total value of shipments from 1963 to 1970 or 1972. The proportion

TABLE 2.7. Selected food manufacturing industry groups and industries: Number of companies and establishments, 1972, 1967, and 1963

Industry group and industries	Companies			Establishments		
	1972	1967	1963	1972	1967	1963
	number					
Meat products:						
Meat-packing plants	2,291	2,529	2,833	2,474	2,697	2,992
Prepared meats	1,207	1,294	1,273	1,311	1,374	1,341
Poultry-dressing plants	407	709	842	652	843	967
Dairy products:						
Fluid milk	2,025	2,988	4,030	2,507	3,481	4,619
Cheese, natural and processed	739	891	982	872	1,026	1,138
Ice cream and frozen desserts	561	713	901	697	850	1,081
Preserved fruits and vegetables:						
Canned fruits and vegetables	765	930	1,135	1,038	1,223	1,430
Pickles, sauces, salad dressings	429	479	541	495	527	588
Frozen specialties	388	495	566	435	607	650
Grain mill products:						
Prepared feeds	1,579	1,835	2,150	2,120	2,355	2,590
Flour, other grain mill products	340	438	510	457	541	618
Blended and prepared flour	115	126	140	137	148	165
Bakery products:						
Breads, cakes, and related products	2,801	3,445	4,339	3,318	4,042	5,010
Cookies and crackers	257	286	286	315	348	356
Sugar, confectionery products:						
Raw cane sugar	60	61	50	77	83	74
Confectionery products	917	1,091	1,142	1,011	1,183	1,211
Beet sugar	16	15	11	61	65	65
Fats and oils:						
Cottonseed oil mills	74	91	115	115	150	188
Soybean oil mills	54	60	68	94	102	102
Animal and marine fats and oils	402	477	516	511	588	615
Miscellaneous food, kindred products:						
Food preparation	1,856	1,824	1,977	2,099	2,082	2,190
Macaroni and spaghetti	179	190	207	194	205	221

Source: U.S. Bureau of the Census, *Census of Manufactures, 1972, Volume I: Subject and Special Statistics,* U.S. Government Printing Office, Washington, D.C., 1976.

was even higher when an adjustment was made for regional or local markets or census industry definitions.

Value added by manufacturing companies in food and kindred products has been more than double the value added by the farming sector. This has raised questions about the feasibility of cooperatives gaining more of this value. Some managers see this as encouragement for greater vertical integration by cooperatives.

Profits of the food manufacturing industries have not been as great as most manufacturing industries until the 1970s. The percent return on stockholder equity for all food manufacturing ranged from 9 to 11.3 from 1963 to 1972.

Some of the food processing companies are very large. "In the 1970-71 fiscal year, 73 firms primarily engaged in food processing each had assets of over $100 million. Of these, 35 had assets of over 250 million."[44]

By comparison, cooperatives are much smaller. For example, in 1975 the 4 largest cooperatives handling dairy products had assets valued at $723 million. This, however was only 12 percent of the value of assets held by the 4 largest noncooperatives handling dairy products.[45]

TABLE 2.8. Concentration ratios for selected food and beverage manufacturing industries, selected years

Manufacturing industry (classification title)	SIC code number	Value of industry shipments accounted for by the four largest companies							Factors adjusted for:‡
		1972*	1970	1966	1963	1958	1947	1966†	
		percent							
Meat-packing plants	2011	22	23	27	31	34	41	(40)	R
Sausages and prepared meats	2013	19	16	16	16	..	..	(30–40)	R
Poultry-dressing plants	2015	..	16	17	14	..	..	(20–30)	R
Creamery butter	2021	45	s§	s§	11	11	18	(50)	R
Cheese, natural and processed	2022	42	43	44	44	..	..	(60)	R, B
Condensed evaporated milk	2023	39	39	45	40	..	..	(50)	R, B
Ice cream and frozen desserts	2024	29	30	33	37	38	40	(70)	L
Fluid milk	2026	18	20	23	23	23	22	(60)	L
Canned cured seafoods	2031	..	53	42	38	..	..	..	
Canned specialties	2032	67	66	63	67	..	..	(80)	B
Canned fruits and vegetables	2033	20	21	24	24	..	..	(40)	R, B
Dehydrated food products	2034	33	33	31	37	45	56	(50)	B
Pickles, sauces, salad dressings	2035	33	38	33	36	..	..	33	
Fresh or frozen packaged fish	2036	..	38	33	36	..	..	33	R, B
Frozen fruits and vegetables	2037	29	26	24	24	..	..	(40–50)	R
Flour and grain mill products	2041	33	30	31	35	38	29	(43)	R, B
Prepared feeds	2042	..	24	23	22	22	19	(30–40)	
Cereal preparations	2043	90	90	87	86	83	79	87	
Rice milling	2044	43	50	45	44	43	33	45	
Blended and prepared flour	2045	68	s§	69	70	..	..	(75)	B
Wet corn milling	2046	63	64	67	71	73	77	67	
Breads, cakes, and related products	2051	29	29	25	23	22	16	(50)	R, B
Cookies and crackers	2052	59	59	59	59	..	..	(70)	R
Raw cane sugar	2061	44	45	50	47	..	..	50	
Cane sugar refining	2062	59	59	63	63	..	..	(40)	N
Beet sugar	2063	66	65	68	66	64	68	(40)	N
Confectionery products	2071	32	30	24	15	18	17	(40–50)	B
Chocolate and cocoa products	2072	74	79	78	75	71	68	(83)	B
Chewing gum	2073	87	85	88	90	89	70	88	
Malt liquors	2082	52	46	29	34	28	21	(65)	R
Malt	2083	48	s§	41	38	50	49	41	
Wines, brandies, and brandy spirits	2084	53	s§	41	44	35	26	41	B, I

TABLE 2.8. (Continued)

Manufacturing industry (classification title)	SIC code number	Value of industry shipments accounted for by the four largest companies							Factors adjusted for:‡
		1972*	1970	1966	1963	1958	1947	1966†	
		percent							
Distilled liquor, ex brandies	2085	47	47	55	58	60	75	55	B, I
Bottled and canned soft drinks	2086	14	13	14	12	11	10	(30–40)	L
Flavoring extracts and syrups	2087	66	61	63	62	55	50	63	
Cottonseed oil mills	2091	43	39	38	41	42	43	...	
Soybean oil mills	2092	54	56	57	50	40	44	57	
Vegetable oil mills	2093	70	60	53	38	66		53	
Animal and marine fats and oils	2094	28	s§	25	23	20		25	
Roasted coffee	2095	65	58	54	52	...	...	(50–60)	B
Shortening and cooking oils	2096	...	46	47	42	44	...	(80–90)	L
Macaroni and spaghetti	2098	38	s§	31	31	25	23	31	
Food preparation	2099	26	26	26	24	...	...	(60–70)	B
Cigarettes	2111	84	84	81	80	79	90	81	
Cigars	2121	56	61	58	59	54	41	58	
Chewing and smoking tobaccos	2131	71	s§	59	58	57	61	59	B
Tobacco steaming and redrying	2141	67	66	69	70	73	88	(70–80)	B
Weaving mills, cotton	2211	31	33	30	30	25	...	(40)	B
Weaving and finishing mills, wool	2231	38	54	56	51	...	...	(50)	B
Leather and sheep-lined clothing	2386	...	s§	24	28	18	24	(60)	B

Source: Alex F. McCalla and Harold O. Carter, "Alternative Agricultural and Food Policy Directions for the U.S.: With Emphasis on a Market-Oriented Approach," Preliminary report prepared for Policy Research Workshop on Public Agriculture and Food, Price and Income Policy Research, Washington, D.C., Jan. 15–16, 1976.

* U.S. Bureau of the Census, *Census of Manufactures, 1972, Volume I: Subject and Special Statistics*, U.S. Government Printing Office, Washington, D.C., 1976.

† Adjusted concentration ratios.

‡ R—markets are mainly regional in scope.
L—markets are mainly local in scope.
B—census industry definitions are too broad.
N—census industry definitions are too narrow.
I—imports are a significant fraction of total sales.

§§—Data suppressed because some of the largest companies were approximately the same as other companies not included in the sample. Therefore, a reliable numerator could be computed.

53

Farmer Cooperative Grain Company, Haven, Kansas. This is a local marketing cooperative that is a member of FAR-MAR-CO, a regional grain marketing cooperative.

TABLE 2.9. Estimated marketing business of cooperatives, by specified commodities, 1975-76

Farm products marketed	Net volume of business*
	millions
Beans and peas (dry edible)	$ 116
Cotton and cotton products	960
Dairy products	8,480
Fruits and vegetables	2,861
Grain, soybeans, and soybean meal and oil	10,634
Livestock and livestock products	2,784
Nuts	555
Poultry products	807
Rice	869
Sugar products	1,337
Tobacco	291
Miscellaneous	122
Wool and mohair	21
Total	$29,837

Source: Bruce L. Swanson and Jane H. Click, *Statistics of Farmer Cooperatives, 1972-1973, 1973-1974, and 1974-1975,* FCS Research Report 39, FCS, USDA, Washington, D.C., Apr. 1977, p. iv, preliminary.
*Excludes intercooperative business.

In fiscal 1976 cooperatives marketed nearly $30 billion of farm products (Table 2.9). Dollar volume of business was heaviest in grain, soybeans, and soybean meal and oil. Dairy products ranked second in dollar volume of business.

Most marketing cooperatives, except for dairy and fruit and vegetable cooperatives, are not greatly involved in processing. In 1969 cooperatives accounted for 21 percent of the total United States pack of fruits and vegetables.[45]

In 1973 dairy cooperatives processed or manufactured only 28 percent of total milk delivered to plants and dealers.[47] And in 1975 cooperative meat packing accounted for only 1 percent of cattle and 2 percent of hogs slaughtered.[48] Cooperative grain processing (except for the processing of soybean oil and meal) is also very limited. Most cooperative activity is restricted to assembling, storage, and, to some extent, transportation.

The situation was as Martin A. Abrahamsen reported:

> In other words, most marketing cooperatives do comparatively little processing, integrate operations to a minor extent, and achieve only limited market penetration. Put another way, as a group, they do not perform many of the major marketing functions involved in moving farm products from the producer to the consumer. Major exceptions to this observation are the dairy cooperatives that manufacture butter, cheese, and related products and fruit and vegetable processing cooperatives.[49]

Expenditure for research and development (R&D) in the food processing industry has been relatively small in comparison to most other industrial sectors. Most of this has been by the larger companies.[50] According to Greig, most of the growth in food processing has come not from the development of new food products, but from "the development of convenient, processed forms of well-known foods using combinations of well-known, readily available food processing technologies."[51] The failure to develop new food products rests with the fact that it is very costly, and only a few of the large processing firms can afford the large initial investment for development and promotion of new products.

ASSESSMENT. Cooperative leaders view processing and marketing as important for one or more of the following five reasons: (1) to extend the farmer's involvement in his product to enhance his total share of the consumer's dollar; (2) to exploit food research and development applications, particularly to protect the outlet for farm-produced goods, often by development of new uses for farm goods; (3) to assure farmers a direct outlet for their products, even if there is not a ready market; (4) to enhance bargaining strength; and (5) to maintain consumer acceptability of farm-produced goods.

The first objective, "to extend the farmer's involvement in his product to enhance his total share of the consumer's dollar," raises many questions. Does the farmer need this means of investment? Do his public sanctions fully support this, or could they be weakened by these movements? Does the cooperative have a comparative advantage in this activity? Is this objective fully supported by the rank-and-file cooperative membership? What kind of additional market or distribution activities will be necessary to exploit fully this increased activity.

A big issue exists as to whether the cooperative has a management and economic advantage over the noncooperative firm in these functions. Possibly a bigger related issue is whether the cooperative may need to protect the buffer of competition provided by the noncooperative companies in the food processing sector, especially the independent (nonchain) companies. The independent company is vulnerable to these movements. The cooperative can integrate into processing and marketing and into the business of the independent company. The noncooperative retailer can integrate backward into this area. The retailer and cooperative can share the market with both being involved at different levels and both taking business from noncooperative organizations. Regardless of who has the processing, the retailer and the cooperative can contract with each other and possibly foreclose noncooperative businesses. Insofar as the cooperative strategy is one of being an enforcer and moderator of competition, the long-run consequences of eliminating noncooperative firms may be detrimental.

The second objective, "to exploit food research and development applications, particularly to protect the outlet for farm-produced goods, often by development of new uses for the farm goods," raises many of the same questions as the first objective. Of course, the two are related. One important issue with regard to this second objective deals with the level of dedication of the large noncooperatives to their food divisions. Are other divisions in the noncooperative more profitable? Will the noncooperative put the proportional research and development resources into the food division that it did in the past? Is there a risk to the farmer that nonfood goods can be developed so his raw material is no longer needed? Could the conglomerates exploit these goods to the farmer's disadvantage? Is it better to go directly into research and development in individual cooperatives or to coordinate this work by joint ventures among cooperatives? This has been done by the dairy cooperatives and to some extent by citrus cooperatives. The massive capital expenditures for this approach may not be feasible for cooperatives. Farmers have had a history of capital rationing, externally and internally. Will farmers support this much of an investment?

The third objective, "to assure farmers a direct outlet for their products, even if there is not a ready market," is questionable. Some farm markets have been lost by the closing of noncooperative factories. One notable example of where farmers felt they must protect an outlet was in sugar production. Many local cooperatives have been set up to provide a market outlet in a particular area. But cases where a commodity does not promise a profitable outlet for noncooperative firms should be checked carefully. The cooperative should hold most of these ventures suspect as a profitable and appropriate strategy until they can be fully justified as having a specific advantage for the cooperative.

Often a special enterprise participated in by only part of the patrons of a cooperative, such as in poultry production, makes it feasible for the total cooperative to help underwrite a venture in processing and marketing for part of their membership. The equity issue must be handled, and many cooperatives have handled it.

At the macro level, cooperatives may need to enter processing and marketing to preserve the price-making functions of the market. Central market volumes have declined with vertical integration. Also, a special product such as fruits and nuts in a fairly small geographic production area may need processing and marketing to establish or maintain a national market. This could occur for many reasons such as balancing supply with demand; product promotion, product development, and quality control; maintenance of public sanctions; and development of new markets by product line, or development of an international market.

The alleged need for cooperatives to enter processing and marketing just for a direct available market must be checked carefully. Often a poor opportunity for noncooperative firms is a poor opportunity for cooperatives. If the cooperative needs to go this far, a key question then is, Why not follow on through to distribution as well? The special cases of each instance must be assessed in building a competitive strategy.

The philosophy of developing a market simply because something is produced also may be questionable. Farmers are obviously tailoring their production more and more to the market. As cooperatives process and market more of their products, this will be increasingly so. A strategy that is dictated by processing or marketing only what production relationships dictate can be vulnerable. Competition necessitates matching production with demand; strategy must consider both.

The fourth objective of entering processing and marketing "to enhance

bargaining strength" may be germane at either the individual cooperative level or the macro level. Many cooperatives, such as dairy, have gone into processing and marketing activities to enhance bargaining. This has been particularly important in perishable products where seasonal surplus must be handled, and where bargaining is precarious unless a diversion alternative is available.

The whole nature of how bargaining is to be used as a general strategy for cooperatives will dictate the degree to which cooperatives enter processing and marketing. Processing and marketing will often be necessary in order to bargain effectively. This can come in two different organizations as in the example of fruit marketing in California. A bargaining cooperative there established the price, and the processing cooperative guaranteed an outlet. In some instances cooperatives with processing and marketing may be missing a bet by not picking up the complementary bargaining function in its strategy. Another alternative is to engage in a joint venture or merge with a bargaining cooperative.

American Rice, Inc., at Lake Charles, Louisiana, is owned by more than 1,700 Louisiana and Texas rice farmers and markets about 15 percent of the United States rice crop.

A fifth objective for going into processing and marketing "to maintain consumer acceptability of farm-produced goods" identifies general market accessibility as important. This differs some from the second objective where the cooperative is trying only to increase its share of the value added to the product, primarily by new products. Processing and marketing will be necessary functions if consumer acceptability is the prime objective. This can mean a total strategy that is thoroughly integrated, capital intensive, managerially complicated, and much removed from the original production-oriented cooperative. The instances in which cooperatives would enter processing and marketing will be dictated substantially by local marketing and production characteristics of the cooperative and its patrons.

Fewer and fewer noncooperative firms see processing as an effective and economical specialized function. Most are integrated or conglomerated. Many cooperative managers view the marketing function in the same way. Looking to the future, most see marketing as a specialized, low capital-intensive activity such as brokerage or as a facilitating function such as in centralized assembly, display preparation, and temporary storage for the retailer. Few see marketing activities as a specialized business except in this light. Basically, they see both processing and marketing as part of the large completely integrated strategy to get consumer acceptability. If they make money on these functions, fine; but the basic objective is to have direct access to consumers so they can move a volume of products.

A central issue in objective five is timing. Cooperatives must ask themselves seriously if they are getting into processing and marketing as a somewhat specialized activity at the very time that it is unpredictable. Noncooperative companies typically are not now entering this field of specialized units. They are selling off their existing specialized units (e.g., liquidation of fluid milk divisions by conglomerate food companies). They see these units as a part of a total package but not as stand-alone units. In their long-range planning, they place even less emphasis on specialized units. Are cooperatives getting into this activity at the wrong time?

Food Distributors

The food distribution industry is described in various ways and includes several types of institutions. Emphasis in this section is on the retail and food service industries, the two that control entree to the ultimate consumer. These are of major concern to the cooperative strategist.

In some situations, however, food brokers and wholesalers can be important institutions in cooperative strategy, and cooperative leaders should keep them in mind. Brokers are independent sales agents for the processor.[52] They take neither possession nor title to products. Increasingly, processors sell to wholesalers and retailers through food brokers rather than through their own sales force. Apart from selling, the broker's principal job is merchandising the products he represents at retail.

Wholesalers link food processors with retailers and institutional outlets. Grocery wholesalers have been classified into three types: (1) retailer-owned cooperative wholesalers, (2) voluntary group wholesalers sponsoring retail groups, and (3) independent wholesalers whose customers have no affiliation with the wholesaler. General line wholesalers handle all products. The largest number of wholesalers are those of a more specialized nature—the important ones being in meat and meat products, fresh fruits and vegetables, dairy products, and poultry and poultry products.

Managers should also be aware of the various kinds of stores. Specialty food stores specialize along lines such as meat markets, vegetable stores, and bakeries. Food discount stores are of two types: best known is the food department or the supermarket in a general merchandise store; and the other is the supermarket that merchandises products on a discount basis. Ordinarily, the supermarket reduces costs by eliminating trading stamps, cutting service to the minimum, and competing on price. The convenience store features convenience of location, quick service, and long store hours. With these attributes, it competes against the supermarket even though it has fewer brands and usually higher prices.

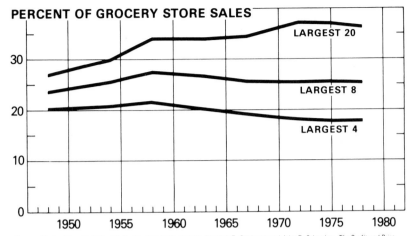

Source: Marion, Bruce W., Willard F. Mueller, Ronald W. Cotterill, Fredrick E. Geithman, and John R. Schmelzer, *The Profit and Price Performance of Leading Food Chains, 1970–74*, A study prepared for the use of Joint Economic Committee, Congress of the United States, U.S. Gov't. Printing Office, Wash., D.C., April 12, 1977, p. 11 and *Weekly Digest*, June 24, 1978, p. 8.

FIG. 2.1.   Market share trends of the 4, 8, and 20 largest grocery chains, 1948–78.

Other specialized institutions exist in food distribution, but most need serious consideration by cooperative strategists only in special cases.

Food retailing is one of the largest industries in the United States. Retail food stores account for about 20 percent of all retail sales. Distinguishing characteristics of the retail food industry are the chain store and the large supermarket.[53]

The largest 4 grocery chains' share of grocery store sales has been declining (Fig. 2.1). The largest 8 chains' share of grocery store sales has leveled off from a decline and is around 25 percent. The largest 20 chains accounted for 37.1 and 36.2 percent of grocery store sales in 1972 and 1978 respectively.

In 1976 the largest 50 nonpublic and publicly held grocery chains accounted for 41 percent of total United States grocery sales (Appendix, Table A.5). Concentration of sales by firms in some local markets, however, have increased, and the concentration is often quite high (Table 2.10). In 194 metropolitan areas in 1972, an average of 52 percent of grocery sales was accounted for by the 4 largest grocery chains. In several regions, the Washington metropolitan area for example, the top 4 grocery store companies accounted for 70 percent or more of grocery store sales.[54]

The supermarket evolved in the 1930s and combined self-service and cash-and-carry merchandising with the centralized food department. It was well suited to the suburban living trends following World War II. The supermarket was developed by independent food retailers, but food chains quickly adopted it. Also, independent retailers combined into affiliated units. As a result, the primary structural units in food retailing are chains and affiliated supermarkets. About 25 percent of the total food distributed

62

TABLE 2.10. Share of grocery store sales by four largest chains in the twenty largest U.S. standard metropolitan statistical areas, census years 1954–72

Metropolitan areas ranked by 1970 population	1954	1958	1963	1967	1972*	Percent change 1954–72
New York	41	37	34	33	31	−10
Los Angeles	30	25	30	28	36	6
Chicago	49	52	52	54	57	8
Philadelphia	53	60	61	60	54	1
Detroit	38	50	52	49	50	12
San Francisco-Oakland	27	29	33	40	41	14
Washington, D.C.	56	60	67	70	76	20
Boston	56	48	50	47	49	−7
Pittsburgh	45	53	52	45	43	−2
St. Louis	35	43	43	39	46	11
Baltimore	48	50	54	55	57	9
Cleveland	51	53	56	58	52	1
Houston	35	33	35	32	35	0
Newark	53	48	40	42	44	−9
Minn.-St. Paul	31	38	39	44	42	11
Dallas	53	47	46	42	†	−11
Seattle-Everett	40	38	41	42	49	9
Anaheim-Santa Ana	...	...	...	...	...	...
Garden Grove	40	47	43	39	44	4
Milwaukee	43	47	40	32	57	14
Atlanta	54	56	60	60	55	1
Simple average	44	46	46	46	48	4

Source: Russell C. Parker, *The Status of Competition in the Food Manufacturing and Food Retailing Industries,* N.C. Project 117, WP-6, Aug. 1976, p. 36.
* Preliminary.
† The Census merged the Dallas and Fort Worth SMSAs in collecting data for the *1972 Census of Retail Trade.* The four-firm concentration ratio for the combined SMSAs was 47 percent in 1972. The Dallas-Fort Worth concentration ratio was not used in computing the average for 1972 which was adjusted to be comparable to the 1967 average.

was accounted for by supermarkets in 1948, but within ten years the percentage was nearly 70 percent; in 1974 it was nearly 72 percent.[55]

The away-from-home eating areas were the last strongholds of independent food retailers, and this area has undergone dramatic changes since the mid-1960s. Today, about one meal in three is eaten away from home.

Three broad institutions in the food service industry are "(1) public eating places and institutions serving food to consumers, (2) institutional wholesalers, and (3) food manufacturers and processors that prepare products especially for this market."[56] The industry is composed of more than 300,000 establishments, mostly eating places, and uses a range of products. Many of the particular individual establishments directly serving the customer have not grown much in size, but they have been consolidated into a large distribution system. Many institutional innovations have been designed to move increasing amounts of food through the various public eating places and institutions. Food producers, processors, and distributers today must consider this large, fast-growing industry in any strategic planning (Table 2.11).

ASSESSMENT. *Abundance and affluence have great effects on food distribution.* Historically, food distribution largely was geared to the con-

TABLE 2.11.   Estimated food service sales and number of units, 1976 and 1980

Institutions	Sales		Units	
	1976	1980	1976	1980
	million dollars		*number*	
Restaurants	28,490	40,640	173,150	169,690
Fast foods	15,240	25,740	104,990	140,210
Retail	2,560	3,750	55,000	59,000
Colleges	3,063	3,843	2,830	3,060
Schools	6,600	8,332	112,700	109,000
Health	6,742	9,378	28,660	29,082
Employee	4,455	5,875	...	...
Transportation	931	1,267	...	...
Hotel-motel	2,990	4,380	43,200	43,000
Recreation	1,890	2,970	25,400	23,500

Source: J. David Morrissy, "Opportunity for Cooperative Growth: Food Service Industry," *Farmer Cooperatives,* Vol. 43, No. 6, FCS, USDA, Washington, D.C., Sept. 1976, p. 5.

cept of scarcity and efficiency.[57] However, food distribution has adjusted to growing abundance and affluence, and the food service industry is a significant adjustment in this direction. The modern chain supermarket has also changed, offering a large number of nonfood items and a tremendous amount of service. A big issue for cooperatives is whether or not they can diversify into this type of market and compete effectively with the skilled merchandisers now selling to today's discriminating consumer. The consumer wants not only a product but a product furnished in a specific way and at a specific time with specific services.

Market power of the retailer is a key evolving issue of the food industry. Probably the most significant finding of the National Commission on Food Marketing was the following:

> Market power is the ability to influence prices or other terms of trade in a way favorable to the business firm. It may be gained through a firm's own strong position or conferred upon the firm by the weakness of those with which it deals. Two groups in the food industry appear to have substantial market power: retailers, including many of the small chains; and large manufacturers, usually diversified, with strong national brands.[58]

Retail structure is important to cooperatives at the national level for two reasons. First, the sheer competitive power of the retailer can be overwhelming. Coordination and concerted interaction may be loosely formulated over a series of local and national markets. But the retailer is recognized and vital in United States food distribution. The size of the retail subsector and interdependence within the subsector give it strength. Second, most product procurement by retailers is concentrated at the regional and national centers where large buyers represent chains and affiliated independents.

Retail structure is important to cooperatives at the local and regional level for two reasons. First, the regional and national procurement program of retailers, mentioned above, can put a local or even a regional cooperative at a competitive disadvantage. Second, the probable entry level for most cooperatives is at the local or regional level. Here they should expect to engage substantial sales competition and concentrated markets and may have to use consolidated power of their own.

Erosion of retailer power has been occurring in two ways. First, a move toward centralized functions by products exists. These tend to take the form of highly specialized functions of wholesaling and merchandising. The ultimate goal here is a store that is merely a framework into which palletized, prearranged displays can be wheeled into place each morning. This store is literally rebuilt at the centralized warehouse each day. This type of functional concentration may bring a whole new competitive structure in procurement. A fundamental question for cooperatives lies in whether this is a good place for them to enter.

The second area of erosion of retailer power has been in the food service industry. Consumer outlets are many and often small, but the buying units serving them are often large. The food service industry has been a fast-growing industry. Many variations exist; it is a complicated and difficult industry to service. Food store retailers are competing with this unit by installing various consumer outlets and services, such as the delicatessen, within their own stores.

Market power of retailers at the local level is often highly concentrated, even when retailer power at the national level may not be highly concentrated.

Interests of the farmer and the retailer are diverging. Cooperatives must consider this trend. Substantial arguments have been made that retailers, because of their merchandising program and the number of non-farm oriented products they handle, fail to secure the highest prices for farm-produced products. In this sense, cooperatives have a significant decision to make with regard to how the retailer fits into national farm income and price policy.

The highly diversified product line of the retailer means that new products must be continually fed into the line. Great difficulty arises in determining how new product choices are made. Moreover, the increasing competition of nonfood items is well established. The products of cooperatives, especially specialized ones, are of lesser and lesser specific influence on the traditional diversified retailer.

The pricing policy of retailers is basically one of obtaining an aggregate average margin. This may include sale of items at a low margin or loss. This can substantially impact the price of individual commodities. Cooperatives, especially those that are commodity specific, have a special problem with these merchandising practices. For example, retailers can use broilers as price leaders and make money on their total operation, but questions arise as to whether the broiler industry is helped.

Accessibility to the consumer is probably the biggest issue for cooperatives with regard to the food distribution system. Noncooperative food processors, manufacturers, and distributors consider this of paramount importance. Most important accesses to the consumer today are through the large supermarket chain and the food service wholesaler. Both of these are difficult areas for cooperatives to enter. Noncooperative food processors and distributors at the national level usually have made long-range plans that guarantee them not only accessibility to the food retailer but to the food service industry as well. They believe it is imperative that they get a share of the substantially increasing business of the food service industry, and to do this they must acquire a direct entree to an outstanding wholesaler who has access to the food service subsector.

The retailer has been the farmer's basic representative to the consumer. A fundamental decision rests as to whether farmers can continue to depend on the retailer for that function. Service has become an important component of consumer needs. Price and service are tied together in the retailer and food service industries. A key question for cooperatives is whether they can move into a food industry so geared to services and compete with the established specialized merchandisers. This is an area quite removed from the raw material production with which cooperatives have been associated, and the area is complicated and expensive. It is fast changing and one in which market demand estimates are difficult to make.

Farmers cannot ignore their public relations role with regard to food distribution. The retailer has had the basic focus of consumer wrath about the level of food prices. This has shifted somewhat; and the wrath is now directed more and more on farmers as well as on retailers. Consumers' concern over high food prices makes consumers sensitive to the effect of food exports on domestic prices. Food price policy is an international issue. Consumers in the United States will focus on their domestic food prices, and a serious question exists as to whether agricultural cooperatives can depend on public attitude and policy to protect their image with regard to food prices. This is especially true in a free-market, export-oriented farm policy. They may need to enter the field in order to get more direct entree to consumers. This could put them in a much stronger position to deal with consumers directly, rather than through an intermediary representative, the retailer.

Integration of the food distribution sector, both backward from the retailer into the manufacturing sector and forward from manufacturing into retailing, is of concern for cooperatives. A significant effect is whether or not the noncooperative type of processing organization can exist with this type of integration. An alternative to integration is for the food retailers to contract directly with cooperatives and ignore the current type of middleman processor and distributor. Cooperatives could also integrate into retailing and manufacturing at the same time.

Thus the retailer and food service industries are no longer passive institutions handling the farmer's product. They are tied closely to the farmer—sometimes for mutual benefit, sometimes for competitive frictions. Cooperative strategists must sort this out if cooperatives are to survive.

3 Nature of Cooperatives

THE ENVIRONMENT for farmers and their cooperatives will continue to change. How well cooperatives adjust to this changing environment may well decide whether they survive. Much depends on whether cooperatives help to shape the environment, react to the environment, or do nothing to change with the environment.

Cooperative organizations have been diminishing in number but increasing in size. In 1950-51, the United States had 10,064 cooperatives in marketing, farm supply, and related services; in 1975-76, the total had declined to 7,535. Net volume of business, $8.1 billion in 1950-51, increased to $40.1 billion in 1975-76—a 6.6 percent compounded annual growth rate.[1] However, cooperative net business volume in 1974-75 was $42.3 billion. The 1975-76 business volume was the first such decline since 1954.

A large percentage of the sales volume is handled by a small percentage of cooperatives. In 1972-73, 240 regional cooperatives—3.1 percent of total United States cooperatives—handled 66 percent of total cooperative volume. At the same time, 96 percent of all local cooperatives handled only 22 percent of total cooperative business volume.

In 1973-74, all regional cooperatives—577 or 7.4 percent of total United States cooperatives—accounted for 50 percent of the marketing and supply net volume of business handled by cooperatives. In 1975-76, the number of regional cooperatives declined to 541.[2]

Local cooperatives are becoming more dependent on regional cooperatives. Management expertise, financing, and many kinds of services are provided by regional cooperatives. Hence, managers and directors in the regional cooperatives provide leadership for improving the operation of local cooperatives.

Issues and problems of the more progressive or unique type of regional cooperatives were the focus of this study. Implications for both the individual cooperative and cooperatives in general were considered. Managers and directors have difficulty seeing their organization as a part of the over-all picture because each organization has its own specific problems.

Growth of some superlarge cooperatives, for example, has become a controversial issue. Internal problems in such areas as capital, management, and member relations are not the only cause; external problems such as antitrust and food and nutrition policy also bear on the growth issue.

Before cooperative leaders can solve problems facing their organizations, they need to have a clear understanding of cooperatives' purposes or reasons for being. Objectives can then be more clearly defined, and general strategies to fulfill cooperative objectives can be selected.

Too often, concern of managers and directors with day-to-day operations overshadows thoughts about long-run objectives and goals. Day-to-day operations without some clear objective or long-range goals can lead cooperatives astray.

Reasons for Being

Cooperative managers, leaders, and policymakers must continually scrutinize their reasons for being. Lack of knowledge as to why cooperatives exist today can result in poor decision making.

Cooperative managers, directors, and leaders must answer some important questions regarding existence and continuance of cooperatives. Some of these are: What is the purpose of your organization? Is this true for cooperatives in general? Have cooperative purposes changed over time? Is your organization and are cooperatives in general straying from their original purposes? Are the reasons that cooperatives exist today in line with the intent of the Capper-Volstead Act? Do cooperatives have more than just a micro-oriented reason for being or should cooperatives be considering more macro-oriented reasons?

THE FORMATION OF COOPERATIVES. Cooperation among American farmers had its beginning in colonial days. At that time farmers helped each other and worked together jointly on an informal basis. Through the years, these efforts became more formal and business cooperatives had their beginning.

The first stirring of the United States cooperative movement was not by organizations formed by agricultural producers. The cooperative movement was brought about by trade workers' dissatisfaction with low wages and poor working conditions. Through cooperation, workers found they could improve conditions. The strike was their main weapon, and by it, they obtained power to negotiate for a minimum wage and a closed shop.

According to Edwin G. Nourse, the first organized movement of agricultural producers began in 1858. At that time, a group of Illinois farmers unhappy with economic conditions, especially monopolistic control by middlemen, met at Centralia to determine what could be done. The farmers agreed that they needed to assert their supremacy, that they along with the purchasers should have more voice than the nonproducers in price determination; that producers and consumers should be brought closer together to diminish the power of nonproducers; and that, in joining together, they would have strength to cope with the existing conditions. Farmers began to organize "Farmers Clubs" to act on their behalf.[3]

During the Civil War the need for agricultural production increased. This resulted in a general higher level of agricultural prices. With higher prices, economic dissatisfaction of farmers as well as the need for cooperation diminished.

In the late 1860s farmers again found themselves in economic distress. During this time the Grange, a general farm organization, became very active in promoting cooperative enterprises.

These cooperative organizations that sprang up often lacked a satisfactory operating strategy. As a result, many did not survive. It was not until 1875, at the annual convention of the National Grange, that the "Rochdale principles" were adopted as a guide for organization and operation of cooperative enterprises.[4] This gave direction to cooperation.

After the Grange, other general farm organizations such as the Farmers Alliance, Farmers Union, the American Society of Equity, and the American Farm Bureau Federation became active in promoting and establishing cooperatives. The major reason for this activity was to provide economic benefits for members. Leaders in the farm organizations realized that only through cooperation could farmers purchase supplies more economically and sell their products at their economic value.

Since the beginnings of cooperation, many cooperatives have come and gone. All of them in some way were organized to enhance the economic welfare of their farmer members. Other reasons for their formation were secondary.

ORIGINAL STATED PURPOSES. Statutes providing for the formation of agricultural cooperatives in the United States had their beginning around the same time that trade workers began forming unions. Nourse reports that the first law regarding cooperatives was passed in Michigan in 1865. Its major purpose was to authorize the organization of cooperative stores. Massachusetts passed a cooperative law in 1866 which spelled out cooperative procedures in more detail than the Michigan law.[5] In 1887 Wisconsin, Kansas, and Pennsylvania enacted cooperative laws.

Although cooperative activity increased during the latter half of the 1800s, "it was not until 1911 that farmers really took the lead in any widespread movement to secure laws under which such organizations could be most advantageously set up and operated."[6] By 1919 nearly every state in the Union had adopted statutes authorizing cooperative organizations.[7]

These statutes required that persons interested in organizing a cooperative file written Articles of Association or Incorporation. In these articles, a statement of the purposes had to be included.

Analysis of Some Cooperative Purposes. Cooperative purposes[8] of the core sample's original stated purposes were analyzed and compared to their current. cooperative purposes.[9] Major interest was in the purposes stated by each organization for providing patrons various services.

In most cases the original purpose statements included rather specific detail on activities the cooperative would provide for members.

Six organizations stated that they would operate on a cooperative basis. Four organizations stated that their purpose was to provide more economical ways of purchasing and distributing for members. Other purposes such as promoting the general welfare of agriculture or providing orderly marketing were not mentioned as frequently. Thus a mixture of purposes was given in the core sample's Articles of Incorporation (Appendix, Table A.6).

In general, the original purposes of the core sample cooperatives stated that they were to engage in activities desired by members and patrons. These activities were to be carried out on a cooperative basis for the economic benefit of members, patrons, and the general welfare of agriculture.

The exact meaning of "activities desired by members and patrons," and "operating on a cooperative basis," was not well defined. The organizations, in practice, generally adopted several Rochdale principles with some individual variation.

The line dividing a cooperative from a corporation was difficult to define. Even Nourse, in 1927, stated:

[Are we] . . . to regard co-operation merely as a somewhat different way of conducting the affairs of a corporation, the precise terms being left to each company to determine for itself and to modify from time to time as it sees fit; or whether certain definite standards shall be set up to embody certain specific doctrines of co-operation as a distinct method of conducting business, and whether these must be made compulsory in order to keep the line between the two business forms clear.[10]

CURRENT STATED PURPOSES. Today, many cooperatives have a written statement of purposes (besides those included in their Articles of Incorporation or Association) approved by their respective board of directors. In the core sample, purposes indicated variation in what managers and directors believe their cooperative should be doing (Appendix, Table A.7). Frankly, present organizational objectives lack cohesiveness and direction.

One common thread among the cooperatives was the stated purpose to operate for the improvement of the economic position of members. This major purpose at the time of formation had continued without change.

The ways and means of accomplishing this major objective, however, may have changed. It was not clear that cooperative leaders were facing this possible new slant, or totally new question, about reasons for being.

"To operate for the improvement of the economic position of members" may not be enough to guide cooperative leaders and policymakers in decision making today. Secondary or more specific reasons for being may be necessary.

In striving to accomplish one major objective, cooperative leaders possibly destroy their chances of accomplishing other important objectives or subobjectives. For example, cooperatives that continue operating without allocating earnings to member-patrons (by paying the corporate income tax) may run the risk of losing the special treatment given to farmer cooperatives.

ASSESSMENT. Comparison of the core sample cooperatives' original stated purposes with current stated purposes leads to the following conclusions:

1. Operating on a cooperative basis (using a variation of the Rochdale principles as a guide) has become less important.

2. Marketing activities have become more prominent.

3. The welfare of employees has become more important.

4. Ownership and control by members are no longer taken for granted.

5. More concern is being given coordination of activities within and among cooperatives.

6. More emphasis has been placed on quality products and inputs, financing, management, and planning.

Current and Future Objectives—Regional Management and Academic Views

Cooperative leaders have a number of objectives to consider in decision making. They are to:

1. Provide patrons essential supplies, services, and processing facilities

2. Help maintain the United States family farm type of system

3. Provide farmers a competitive market environment

4. Represent members on group issues

The above specific cooperative objectives all relate to the more general objective of improving the economic welfare of members. Regional cooperative managers had a wide variation of ideas on the role of cooperatives and what cooperatives need to do to accomplish their objectives.

IMPROVE ECONOMIC WELFARE. Can cooperatives improve the economic welfare of their members? In the future, according to one manager, this may be the real question.

As a whole, cooperative managers did not think the objective of their cooperative was to improve the general farm income. One manager put it this way, "I guess that I justify our existence to some extent by the fact that we do have a fairly big impact on producers whether they use us or not. I don't necessarily like it, but I think part of the justification for our existence is that we do raise the general level of service of regular people on this market and prices too."

Many managers expressed their belief that cooperatives should be run with business practices and principles just like noncooperative businesses, and that cooperatives should not be thought of as some type of religion (that is, that the cooperative can do everything for its members) if they are to be competitive. "No question about that," was the reply of a marketing cooperative manager, "[you] cannot justify a cooperative unless it can return more money to the grower than a proprietary firm." "You have to perform," stated a manager of a supply/marketing cooperative, "The cooperative concept is fine but you have got to deliver. I think that the religion involved can't do anything but die."

Managers of the regional cooperatives as well as the academic people believed that agricultural cooperatives in general, over the past five years, have been managed and operated more like noncooperative businesses than previously. And they believed that if these cooperatives are going to do the job they need to do, they will move further in this direction over the next decade.

One important cooperative principle is the return of net margins to member-patrons according to patronage. Cooperatives directly enhance farmer-patrons' income by allocating a cash portion of net earnings or net margins. In a competitive situation, farmers would receive the same price for their raw product whether they sold it to a cooperative or noncooperative. Consequently, the farmer belonging to a cooperative where net margins are distributed would be better off than the farmer selling to the noncooperative.

Because of increased capital requirements, some cooperatives are finding it more and more difficult to allocate a large percentage of net margins or net earnings to members in cash. Managers of one cooperative believed they could better serve members over the long run by paying taxes on a portion of net margins. In this way, the reserves can be plowed back into the business, and no new earnings will ever be needed to revolve out old member certificates of equity.

The services that the cooperative offers its members are all not necessarily based on net margins to the cooperative. In some cases management has found it necessary to engage in activities for its members on a break-even basis or at a loss. Overall, however, this type of activity should be a well-considered exception to maintain the long-run financial stability of the cooperative.

Younger managers are more concerned than older managers about operating all areas "in the black." Thus some managers expressed the philosophy that if it is not profitable, they would prefer not to engage in the activity. Some managers were also concerned about what the cooperative could offer the farmer in the long run and the cooperative's return on capital. Without a reasonable return on capital, they believed they would have trouble securing external capital. Some balance between what a cooperative offers its members and return on capital must be maintained.

This whole matter does raise some important questions about cooperatives in the area of patronage refunds. If a cooperative organization chooses to pay taxes on net margins, and thereby believes it is more able to service its members, is that organization still fulfilling its major reason for being? Is it more important that the cooperative invest for future generations, or should it operate so as to return net margins earned by its current members? Managers agreed that there should be a balance between the two, but much disagreement exists among managers as to where an equitable balance can be found.

Another important issue related to the economic welfare of cooperative members involves the level of diversification. How diversified should cooperatives become? Should a cooperative handle nonagricultural products along with the agricultural products if these products provide the organization increased earnings? This is an especially difficult question if the earnings can be returned to members or they can be used to provide more and improved services, thereby enhancing the income of its members.

Most managers stated that they could see handling a nonagricultural product only if it is in some way or another member related. They believed that cooperatives should not expand into areas where no farmer-producer ties exist. However, a few managers were not quite as explicit on this issue. Thus future cooperatives may well handle more products that are not member related. What implications this has for the whole cooperative movement is not yet known. It is, however, important that boards of directors and managers review what is best for the economic welfare not only of their members but for cooperatives in general.

PROVIDE ESSENTIAL SUPPLIES, SERVICES, AND PROCESSING FACILITIES. "Cooperatives are uniquely structured to service farm people," stated one cooperative manager. Servicing the producer clientele was mentioned by several managers as the basic justification for the existence of their cooperatives.

"Where you do fulfill a need, they will stick by you" was the general feeling of management of one organization toward its farmer members. According to some managers, there is some loyalty of members to their cooperative, but management cannot depend on this as a major basis for running its business. In the end, it all "boils down to the dollar." Producers like to shop around; if supplies are cheaper from a noncooperative organization, they will often go there. If times get tough, growers come to the cooperative for help and support. Growers look to their cooperative as a "protective umbrella," as one manager put it.

The fertilizer shortage in 1973 was a good example of cooperatives protecting their members. Managers involved in handling fertilizer were proud of their record over the past few years in providing such an essential production item to their members. However, several managers believed that too many farmers take their cooperative for granted. As several cooperative managers put it, farmers soon forget what their cooperative has done for them. Cooperatives do not always gain substantial long-run advantages.

In the future, cooperatives may need to cooperate even more in setting up basic raw material sources so that farmers are provided good quality supplies at reasonable value. Cooperatives will need to do more centralizing of warehouses to reduce inventory and improve distribution. Many managers expressed the need to improve operating efficiency.

Providing processing facilities for members' products will be even more important in the future. If cooperatives are going to become market oriented, they must have the needed processing facilities. This all goes back to one of the original reasons why cooperatives were established—if farmers are going to get more for their products, cooperatives will need to take over some of the functions of the middlemen and put farmers in more direct contact with consumers.

In the past cooperatives have not been very market oriented. Cooperatives can no longer be just a home for the farmers' products. This means changes must be made on the production side. Farmers must produce a certain kind and quality of product. In the future, this may mean that more and more products will be produced under marketing agreements. Those farmers that cannot meet the desired product specifications may find they no longer have a place to market them.

"The thing I'm concerned with," stated one cooperative manager, "[is], if we don't do something to get the farmer more for the product he is producing, we're going to have more and more of our farmers disappearing from the scene." Other managers also believed that cooperatives needed to do more for farmers in the marketing aspects to help them get a better price for their products.

In the selected core sample, all cooperative managers agreed that one of the reasons for the establishment of their organization was to provide essential supplies, services, and processing facilities. This, they all agreed, is true at present and will still be true by 1985. The selected sample of academic people were also in unanimous agreement on this.

A question exists as to how cooperatives, as they get bigger, will know what services farmers need. One manager expressed various ways that member needs come to management: (1) people employed in the field, (2) district meetings, (3) telephone calls or letters from people, and (4) through studies of particular issues. He emphasized the importance of cooperatives keeping close to the farmer and to what is going on. But this extends beyond monitoring of current needs to forecasting what future farmer needs will be.

MAINTAIN THE INDIVIDUAL FAMILY FARMER. When cooperatives were established, concern about the declining number of family farmers was not as it is today. In fact, only half of the core sample respondents believed their organization was originally established to assist in maintaining family type farms. At the present time, and in the future, cooperative managers see an increasing responsibility to protect the family farmer.

The academic community also believed that cooperatives would have more responsibility for maintaining the family type farm in the future. All agreed that agricultural cooperatives as a group would be the key means to maintaining a United States farm production system of many individual farmers.

The family farmer through cooperation can help himself in many ways.

Cooperatives can provide the farmer quality inputs at competitive prices. They can provide services of all kinds that help improve the efficiency of the farm operation. They provide research in some areas to help the farmer increase productivity. And in many cases, they provide a market outlet.

Cooperatives also serve the small farmers, and most managers believed that cooperatives had more of a responsibility in this area than noncooperatives. Management, however, did not believe it was the role of cooperatives to help maintain the small inefficient farmer.

Some cooperatives are tightly knit with a general farm organization while others are completely independent or in conflict. This latter point is especially true where the farm organization is competing with the cooperative. If cooperatives are going to maintain the individual family farmer, a break between the cooperatives and the general farm organizations may be needed or else some well-defined line of responsibility should be agreed upon so that competitive conflicts do not exist.

Several academic personnel believed that some areas needed to be improved if agricultural cooperatives in general were to serve and perpetuate the family-farmer system of commercial agriculture during the upcoming decade. Comments were as follows:

"Cooperatives are the family farmers' only hope, but to adequately service the family farmers they must:

1. Secure member commitment.

2. Increase investment by members to gain ownership of more input sources and to modernize output processing facilities.

3. Upgrade and centralize the marketing functions.

4. Become market oriented—develop new products and go to brand merchandising.

5. Lock together input and output marketing.

6. Upgrade the quality of management."

"More cooperation and coordination among cooperatives" was the reply of one professor, while another thought it "will require new cooperatives designed to fit their needs but probably co-ops *alone* cannot perpetuate them [the family farmers]."

Another professor summed it up, "There is and will be a never-ending struggle for cooperative management to remember their legitimate goals and for cooperative boards to keep the cooperative working for farmers as much as for the employees and managers. The cooperative must be 'good' for all three groups: Farmers, managers, and [other] employees. Many cooperatives do not yet perceive the important roles that they need to play in the survival of their members."

PROVIDE A COMPETITIVE MARKET ENVIRONMENT. Cooperative managers in the selected core sample strongly agreed that their organizations were originally established to provide a competitive market environment. They all agreed that this is one purpose of cooperatives at the present time and for the future.

Cooperatives help to provide a competitive environment if through their actions they prevent themselves or others from monopolizing or taking advantage of the particular situation at hand.

Today's competitive situation for the farmer has not changed much in comparison to the early days of cooperation. Only today's middlemen—the suppliers and handlers—are larger, and raw products may go through more stages of manufacturing and marketing before they reach their final destination. Some managers believed it important for cooperatives to be large enough to compete at all stages of manufacturing and marketing.

The core sample managers were in fairly strong agreement that agricultural cooperatives have to maintain countervailing power in the following stages or areas: (1) at the first assembly or sales from the farm, (2) in finished farm or food products, (3) in the sale of farm supplies, and (4) in the availability and quantity of farm services. The academic people also agreed to the above, but their beliefs were not stated as strongly as those expressed by managers.

According to cooperative managers, farmers do need the umbrella of cooperatives "to keep the corporations honest." A manager of a large regional said, "Members are afraid that if we aren't there as a competitive force they would suffer in the long run." But in many cases, farmers do not realize how important cooperatives are in maintaining a competitive environment.

The significant question is, "Will cooperatives be able to meet the objective of providing a competitive market?" Can cooperatives really compete against the conglomerates without becoming conglomerates themselves?

REPRESENT MEMBERS ON GROUP ISSUES. Cooperatives are also able to influence legislation and represent members on issues that are in the best interest of the family farmer. An individual farmer by himself would have no influence on many of these issues. But is it the responsibility of cooperatives to represent members on group issues?

A majority of the managers in the core sample believed that one cooperative purpose was to represent members on group issues. They did not, however, see this as becoming a more important function in the future. Some cooperative managers believed that this was the role of general farm organizations. Thus these managers did not engage in activities representing farmers when they believed it was the responsibility of the general farm organizations.

Most members of the academic community believed that one purpose of cooperatives was to represent members on group issues.

Cooperative Advantages and Disadvantages

Cooperatives in comparison to public investor-owned corporations have some unique characteristics. This uniqueness does allow for certain advantages and disadvantages. Managers and directors who know these can be more effective in their decision making by doing more in the areas where they have greatest comparative advantage.

Cooperatives are owned by member-patrons. Members own a cooperative by owning membership stock, by investing capital to buy new facilities or expansion, and by allowing the cooperative to keep patronage refunds or per unit capital retains that will be redeemed at a later date.

Corporations are owned by stockholders. Any person—a farmer or nonfarmer—may invest in a corporation that has listed shares of stock for sale on a public stock market. A shareholder may or may not patronize the corporate business.

The differences between cooperatives and noncooperative corporations are listed by Marvin A. Schaars (Appendix, Table A.8). "The differences are primarily in the relationship between the owners and their organization and in the way profits and net savings are distributed."[11]

Cooperatives have a number of distinct *advantages* because they are owned and controlled by member-users.

The member has some say in what activities the cooperative will promote. If a sufficient number of members need a particular service, they can through their own voice have that service provided. These close ties with members mean that the organization has a fairly reliable raw product market. As long as management of the cooperative can provide the necessary services, supplies, or marketing at competitive levels, there will always be some member loyalty.

One of the great advantages of a cooperative is that it can work together with its members and with other cooperatives to market member products. The Capper-Volstead Act of 1922 explicitly sanctions "marketing agencies in common" provided the respective associations conform to certain requirements and they do not through monopolization or restraint of trade in interstate or foreign commerce unduly enhance the price of any agricultural product.[12]

One other advantage that cooperatives have in comparison to noncooperative organizations is access to the Bank for Cooperatives system. The Banks for Cooperatives have provided a source of debt capital that has been extremely important to the operation and growth of cooperatives.

Cooperatives also have some *disadvantages*. Cooperatives do not have one source of capital that is used by corporations, that is, the public sale of voting stock. So far, cooperatives have been able to obtain enough capital through the Banks for Cooperatives and other sources.

The Capper-Volstead Act applies to cooperatives that market agricultural products for members who are agricultural producers. The Capper-Volstead Act does not apply to the marketing of nonagricultural products or the purchasing and sale of farm supplies.

In some cases, agricultural cooperatives need to offer certain services to members, and these services may not pay for themselves. Activities of this nature make it difficult for managers to end the operating year with the type of bottom-line results they would like. Noncooperative businesses do not have this problem to the same degree. If specific activities do not generate a specified return, these activities are usually dropped.

Other limitations faced by agricultural cooperatives are: (1) They are usually required to handle all their members' product even if it means a higher operating cost per unit; (2) detailed records have to be kept for business done by each patron so that net margins can be allocated correctly, and so that tax laws can be complied with; (3) farmer directors are sometimes conservative and not willing to take the risk that is involved in maintaining a progressive operation, and (4) cooperative organizations must keep their members more informed—a costly service for large cooperatives.

Managers reported a number of advantages and disadvantages to the authors. The advantages of cooperatives over noncooperatives as expressed by respondents are listed below but not necessarily in descending order of importance:

1. Access to the Banks for Cooperatives.

2. Availability of special income tax status.

3. Use of patronage refunds—cooperatives often return more dollars to their farmer owners.

4. Grower loyalty in marketing cooperatives—the grower familiarity with the desired quality of product to be produced.

5. Greater working relationship among cooperative members, and the sanction to act together.

6. More open to farmer-members' needs.

7. Local and regional cooperatives are located so that logistically they provide better services.

8. Cooperatives deal directly with members; a noncooperative organization may be a small part of some very large conglomerate whose headquarters and top management are unknown to farmers.

9. Cooperatives have proved that they do not desert their farmer members during times of shortages.

10. Members receive more equitable treatment.

11. Cooperatives are more concerned about developing people.

Stated disadvantages or limitations of cooperatives in comparison with noncooperatives were:

1. Can't acquire capital by offering voting stock.

2. Some commercial bankers hesitate to loan to cooperatives because of bankers' lack of knowledge about cooperatives.

3. In some cases, cooperatives lack unity or uniformity.

4. Planning long-term investments is difficult because growers demand that their returns be good every year.

5. Not as market oriented and lack research and development to keep outlets open for farmers.

6. The one-man one-vote limitation.

7. Some farmers expect the cooperative to offer more services at better prices than anyone else.

8. Limited financial incentives for managers.
9. Limited returns on equity.
10. Limited to the marketing of agriculture-related products.

In summary, cooperatives should not only utilize the advantages they have but also maintain and protect the other advantages mentioned. Some cooperatives are tending more toward the noncooperative form of business in an attempt to capture some of the advantages of noncooperative business. If it is not in the best interest of these organizations to follow cooperative practices and principles and the advantages thereof, it may be in their best interest to convert to the noncooperative form of business. One or two cooperatives operating under principles different from the standard cooperative principle can damage the image for cooperatives in general.

Social Responsibilities

One important question cooperative leaders must answer is, Do agricultural cooperatives, because of their special legal treatment, have a greater social responsibility than do noncooperative organizations? In other words, Should cooperatives be putting more emphasis on maintaining small farmers or other low-income people in rural areas and/or consumers in general?

Most managers agreed that agricultural cooperatives as a group do have greater responsibilities to average farmers and large farmers than do noncooperative businesses. Their thoughts about cooperative responsibilities to small farmers were not expressed as strongly.

A majority of the managers believed that cooperatives had no greater responsibilitiy to low-income people or consumers in general than did noncooperative businesses.

Managers, however, did believe that they had a responsibility to serve the small efficient farmer. But managers did not see that cooperatives had any social responsibility to keep the small inefficient farmer in business.

The responses from the academic people concerning small farmers were about the same as received from managers. However, a majority of the academic people agreed that cooperatives have greater responsibilities to average farmers than they do to large farmers. They did not believe that cooperatives should put greater emphasis on serving the larger farmer than the average farmer.

More academic people than managers thought that agricultural cooperatives had no greater responsibility with regard to consumers in general than did noncooperative businesses. A majority of both groups agreed that agricultural cooperatives were generally meeting their social responsibilities to their members.

In summary, considerable belief existed that agricultural cooperatives have greater responsibilities to farmers than do noncooperative organizations. Cooperatives, however, do not have any greater responsibilites for helping low-income people in rural areas or to consumers in general.

Schaars has written:

A cooperative is not an eleemosynary company despite the fact that it may have a greater social outlook than a standard corporation, nor is it designed to take over the welfare functions of the community or state. These functions are society's responsibilities—not those of the small sector of that society affiliated with cooperatives.

The vast majority of cooperators, especially in the U.S. and, I suspect, in most countries, are chiefly interested in personal economic benefits, lose interest in the cooperative when these are not forthcoming, and have little or no interest whatever in remedying the ills of society.[13]

In a prepared statement for the Graduate Institute of Cooperative Leadership in July 1974, Jim Hightower reported,

The substance of social responsibility in cooperatives is primarily economic. There are those who seem to fear that acceptance of a social responsibility means taking on welfare cases, just as there are those who seem to equate social responsibility with having a hospitality suite at the co-op annual meeting. But it seems to me that the essence of social responsibility in the cooperative movement today is to provide farmers an economic tool with which to fight concentrated, corporate power in our changing food economy. That requires that farmer cooperatives maintain their independence and their organizations' democratic integrity while seeking market strength. . . . The public desperately needs cooperatives to offer an alternative marketing mechanism, an alternative way of doing business. That is an enormous and historic social responsibility.[14]

Components of the Cooperative System

Over the years, cooperatives have developed a system to carry out their reasons for being. The regional cooperatives studied here are only a part of this system. Close to the individual farmer are many local cooperatives that deal directly with the farmer. They may provide a number of services including furnishing supplies, offering of specialized services, and product marketing.

Today, several thousand organizations, "local cooperatives," serve the immediate needs of farmers within a community or county. Many locals have expanded their operations to meet almost all the needs of their members in operating their farm businesses.

Many local cooperatives have formed regional federated cooperatives. Activities of the regional cooperatives may include selling products on central, terminal, or wholesale markets, or wholesaling and manufacturing farm supplies for distribution by the locals.

The federated regional cooperative is to the local cooperative what the local cooperative is to the farmer. Federated cooperatives have local cooperatives as member-owners. Farmers are thus indirectly members of their regional cooperative through their local cooperative. Through the regional cooperative, the local cooperative can buy supplies cheaper and market the farmers' products at a higher price. The regional takes the responsibility of coordinating activities, providing services, and supplying information not only to the locals but directly to members.

Harold Jordan, former general manager of the Indiana Farm Bureau Cooperative Association (IFBCA), told members at an annual meeting in 1968 that advertising, public relations, auditing, computerized data and record processing, research, personnel and management training and development programs, when developed and provided to counties (local cooperatives) on a statewide basis, have been and are proving to be more effective and less costly.[15] Many of the activities that the regional cooperative engages in may depend on its organizational structure—centralized or federated.

A second subsystem of cooperatives consists of regional centralized associations that are owned directly by farmers. These cooperatives serve farmers through outlying branches or facilities. The centralized regional cooperative like the federated regional cooperative may serve a number of counties, a state, or several states. When they become large and well established they often add vertically integrated services such as processing of farm products or manufacturing of feed or fertilizer products.

A few combination centralized and federated regional cooperatives also exist. Part of their members are farmers and part are local cooperatives. They operate some local facilities and also serve local cooperatives.

The regional cooperatives, both centralized and federated, have gone a step further and organized interregional or national cooperatives. These organizations were organized through joint efforts of regionals for the purpose of engaging in activities—procurement, marketing, research, or petroleum production where it would be too risky and/or uneconomical for the regional cooperative to do it alone. More will be said later on this type of multi-cooperative organization.

Consequently, farmers have developed cooperatives to provide services at local, regional, and to a lesser extent across regions or at national levels. The marketing cooperatives perform increasingly complete, integrated services from local assembly to terminal sales, to processing, and to exporting of farm products. The supply cooperatives perform integrated services from manufacturing and producing of raw materials to wholesaling and local retailing.

Growth Patterns and Issues

Growth has been a significant means by which businesses adjust to the environment. Growth is an American tradition. Yet growth can distort the needs for other adjustments, and it can set the stage for great problems once growth stops. Cooperatives have grown, are growing, and will continue to grow according to expectations expressed by managers. The strategic factors of growth are important in cooperative planning.

Some hold that cooperatives are too small to do the competitive job and that cooperatives must grow. Yet some cooperatives that have grown rather rapidly to fairly large businesses, such as the dairy cooperatives, have experienced problems.

Some government officials and private business managers believe that cooperatives have grown too large. They question the right of cooperatives to use special legal treatment to help obtain such size. Some private business managers are saying that this has led to unfair competition.[16]

Many strategic plans revealed by managers suggest the need for much larger cooperatives. Growth requirements for the strategies that are most seriously planned by the more innovative cooperatives raise important questions.

The focus of this section is on such questions as What is growth? How is growth measured? By what methods can cooperatives grow? What are the limitations to cooperative growth? and Is growth an adequate strategy for cooperatives?

Business management experts emphasize the necessity of growth. Frederick E. Webster summarizes it thus:

> Two forces motivate the healthy firm to grow: (1) the need to create new opportunities for owners and employees in the form of new earnings, higher returns on investment, enhanced and more rewarding responsibilities, and the challenge of new problems; (2) the need to adjust to changing market conditions (customer preferences, new competition, and new technology). Because the environment is changing, the healthy firm has no choice but to continue to change with it or face the inevitable consequences of obsolescence—lower profitability, sales, and return on investment; higher operating costs; the need to remove personnel from the payroll; and eventually bankruptcy or liquidation. Preservation of the status quo is simply not a feasible objective for the firm. It must change if it is to survive, and it must grow if it is to meet the expectations of owners and employees, and their needs for personal growth and advancement.[17]

CONCEPTS OF GROWTH. Growth is defined as an "act, process, or manner of growing." Growing is "to increase by natural development"; to "increase in size or substance"; or "to increase in influence or effect."[18]

Many think of business growth as an increase in size. But size can include many things—an increase in number of employees, plants, production, or other operating characteristics. Leon Garoyan defines growth as "any change in size, scale of operation, or capability to serve members or customers which results in profits or in improved services that will bring the desired long-term results."[19] The fact that this definition rules out any increase in size that may not be profitable or that does not improve services appears to be a limitation in this definition.

Richard W. Schermerhorn includes the concept of adjustment in his definition of growth: "a cooperative achieves growth if it is successful in adjusting its operations in line with current business conditions, even though it may not have actually increased its dollar volume of business, or

In 1929, the Union Oil Company Cooperative in North Kansas City, Kansas, was born. By 1979, this cooperative, now known as Farmland Industries, Inc., has become the nation's largest farm cooperative. The building is Farmland's headquarters in Kansas City, Missouri.

added to his [the cooperative's] physical assets, or for that matter, added new members to its rolls.''[20] This concept fits into growth issues of this study.

Most cooperative managers believed that their organization needed to grow in some way—either in terms of increasing their market share or in doing things better than now. "We want growth for the farmer and service for the farmers," stated one manager; such was the general thought of many managers.

Growth, however, has to be planned. It is not a matter of jumping into activities that appear to have great potential for success. One cooperative that had grown at a fast rate and then found itself in financial trouble explained that it had grown "too far too fast." It had overextended itself into many areas without considering what it did best and then expanding that base. An executive said their philosophy at that time was, "We're not sure that this will make money but it will help the existing products." And in some cases that cooperative lost money.

Cooperative managers still consider growth important to cooperative

survival, but they greatly emphasized research and planning before making decisions for adjustment or expansion.

Assigning priorities to growth was expressed by a manager as depending on (1) need, (2) how it strengthens the total organization, and (3) the rate of return. "When the costs go up," explained a manager, "you have to increase the price of your product and you have got to get more products to go with it. Otherwise, you're not going to offset the increased costs. Any time you stop growing, you start sliding backwards."

Fred Koller summarized why cooperatives must grow to survive: "Cooperatives need to grow larger to take advantage of a continuum of new technologies, new opportunities for economies of size, and increased efficiency. Likewise, in view of a rapidly changing overall market structure toward concentration and greater market coordination, cooperatives must stay abreast of these trends to survive."[21]

Managers were looking to the needs of their members, and their organizations were growing to meet these needs in a variety of ways. One manager gave two philosophies about growth: (1) A firm should grow to increase sales, or (2) a firm should maintain a strong foundation so that the organization can grow wherever the opportunity presents itself and this growth can be obtained without disturbing the foundation.

In a merger study, Garoyan and Gail L. Cramer stated, "The need for growth is a recognition of the need to achieve economies of size in plant processing, distribution, and management, to enhance bargaining power, to improve cooperative effectiveness in the marketplace, and to provide a stronger financial base."[22]

MEASUREMENTS OF GROWTH. Cooperative managers used various measures to show growth: change in dollar sales, market share, memberships, new programs, new products, services provided, and physical units handled.

In business literature, the most often used growth measures are based on financial aspects such as sales, assets, net worth, or profits. In the case of cooperatives net earnings or net margins would be used in place of profits.

Nonmonetary measures of growth often used by businesses are changes in number of employees, market share, and physical units. Cooperative business organizations sometimes use changes in number of memberships.

Extent of horizontal and vertical activities may be a measure of growth. In many cases, the growth measure referred to will depend on whether it is used to denote growth in an individual business or growth for the industry as a whole. W. Smith Greig, for example, in a study of the food processing industry, measured growth with changes in value added, employment, value of shipments, and capital expenditures.[23]

According to Joseph W. McGuire, "Net worth and profits . . . reflect company well-being, but have the disadvantage of being compound variables (i.e., dependent upon sales and costs) and may vary because of factors which are not significant to growth." These factors include such items as accounting procedures, depreciation, and inventory.

Asset values would be a fairly good measure of growth for longer periods. They are not a sensitive measure of growth for changes over a short period.[24]

Number of employees is a fairly good measure if a nonmonetary unit of measuring growth is desired. The measure would not be inflated as found *with monetary* units, and increased output is correlated with additional employees. However, this measure would not reflect the increase in growth due to the substitution of capital for labor. There would also be some lack of sensitivity in the short run. Usually, there is a lag between the additional number of employees and output.[25]

Number of memberships and market share are not always good measures of growth for the individual firm. An increase in membership does not necessarily imply a corresponding change in other measures—total sales, for example.

Market share in most cases is difficult to measure, especially in a market of many firms. Also, the measure would not have much sensitivity to growth over a short period.

Managers did want to keep track of market share. In some cases, a minimum market share is set as a goal. In a few cases, managers were worried about increasing their market share because of possible antitrust accusations and investigations brought on by other businesses, the Federal Trade Commission (FTC), and the Justice Department.

Market share and profitability are highly related. In a study of fifty-seven companies, Robert D. Buzzell et al. found that "one of the main determinants of business profitability is market share." Three possible explanations were given: businesses with larger market shares (1) can achieve greater economies of scale; (2) can exert greater market influence and can bargain more effectively; and (3) have managers with greater qualities in leadership abilities.[26]

The most often used measure of growth by businesses is gross sales. McGuire holds it to be the "most effective" measure of growth. Sales show good sensitivity to economic conditions, are fairly easy to calculate, and possibly are more accurate than other measures. The sales measure can be used to portray changes in both short- and long-run periods. However, inflation can cause problems in using a sales measure.

McGuire warns that changes from year to year in actual sales as a measure of growth may not be evidence of growth. "The actual sales of firms respond not only to long-term movements, but also reflect changes caused by short-term economic fluctuations, and are affected by business cycles and random factors." He also states that "Absolute or percentage changes in sales over a ten-year period are also unsatisfactory as a growth measure. In the latter measure all of the weight is placed upon the sales values at the extremes, and in many cases sales in the extreme years may not be typical of sales during the remainder of the period." He recommends that trend analysis or statistical curves be used to quantify growth.[27]

Although change in sales is, perhaps, the most often used measure to indicate growth, an increase in sales does not necessarily imply a corresponding increase in net savings. CENEX (Farmers Union Central Exchange, Inc.), a regional supply cooperative headquartered in St. Paul, Minnesota, illustrates the point. In fiscal 1976 sales were $570.5 million; in fiscal 1977 sales were $645.6 million—a 13 percent increase. Savings in fiscal 1976 were 35.7 million; in fiscal 1977 savings dropped to 27.9 million—a 21.8 percent decrease in gross savings in one year.[28]

The implication is that businesses really cannot rely on only one measure of growth. Management needs continuously to evaluate its sales and intermittently to check its growth in earnings and its return on investment. One cooperative manager was explicit in defining this need when he stated: "In all ventures, you have got to measure investment as well as growth."

In looking at food processors, *Business Week* found that "in their quest for earnings growth, many companies lost sight of the fact that their return on investment was being destroyed."[29] This reinforces the need for managers not to become bogged down in trying to accomplish one set goal while failing to give due consideration to other important factors.

IMPORTANCE OF GROWTH. Individual cooperative growth, as expressed by managers, is important for several reasons. First, growth can provide many cooperatives with a means to carry out their strategy and to meet their objectives and goals. Second, growth often allows cooperatives ways to obtain organizational and operational efficiencies. And third, growth may be an effective method whereby a cooperative can maintain (survive) or improve its position in the environment in which it operates.

The cooperative is a tool for farmers to use in expanding their farm business. Cooperatives are an extension of the farm business, and as farm businesses grow, so should cooperatives.

Growth of cooperatives is required if cooperatives are to provide quality supplies and services at reasonable prices. Growth is required if cooperatives are to become market oriented—to search out markets and supply those markets with products that are satisfying to the customers.

There are various ways that businesses organize their management structure. According to Alfred D. Chandler, who studied the historical development and structure of business organizations, the direction of growth or strategy of a business organization over time does determine its structure. "Unless new structures are developed to meet new administrative needs which result from an expansion of a firm's activities into new areas, functions, or product lines, the technological, financial, and personnel economies of growth and size cannot be realized. Nor can the enlarged resources be employed as profitably as they otherwise might be."[30]

Some of the more progressive cooperatives are changing their internal business organization. Several cooperatives are working toward more decentralized management. Each manager is given the responsibility of a particular area or function. How each manager carries out his responsibilities may depend on given objectives of the organization or expected net margins in the area under concern.

Cooperative growth provides a way to accomplish economies in handling, storing, processing, and distribution. It provides a means of reducing procurement and other service costs. It allows the coordination of activities such as supply purchases and marketing. Through growth, duplication of services and other activities can be eliminated.

Individual cooperatives, according to sales volume, have much room to grow. In fiscal 1975 only 11.2 percent of the total number of cooperatives reported individual sales equal to or greater than $7.5 million. And more than 66 percent (4,990 cooperatives) had individual sales less than $2.5 million.[31]

In comparing the total marketing and supply sales of the ten largest cooperatives with the ten largest comparable noncooperative businesses, Martin A. Abrahamsen[32] in a study involving cooperative growth found the sales of the ten cooperatives were only 13 percent of the total sales of the noncooperative businesses. And "cooperatives' net margins, assets, and net worth were only 4 to 6 percent of those for the other firms."

Abrahamsen also concluded from comparison of the four largest cooperatives with selected noncooperatives that "actual growth of the four largest cooperatives in each of 11 selected categories has been appreciably less during the past 20 years than the growth of the four largest comparable firms. The other businesses, for example, had 7 times greater sales, 22 times greater net assets, 25 times more net worth, and 18 times greater net margins than the cooperatives during this period."

In a recent study, Lyden O'Day made comparisons for 1975 of the four largest cooperatives with the four largest other businesses for seven selected industries—grain, fruits and vegetables, milk and other dairy products, poultry and eggs, commercial feeds, fertilizer and lime, and petroleum products. In six of the industries (data for all of the four largest noncooperative grain firms were not available), total sales, total assets, and net worth of the four largest cooperatives were only 11, 7, and 5 percent of the four largest comparable noncooperative businesses.

Comparison was made of the four largest cooperatives with the four largest comparable noncooperatives for five industries (excluding poultry) for 1970 and 1975. The data show the four largest cooperatives' total sales as a percentage of the four largest other businesses' sales were down while total assets and net worth were greater. The four largest cooperatives' total sales, total assets, and total net worth as a percentage of the four other firms' sales, assets, and net worth were as follows:

Year	Total sales	Total assets	Net worth
1970	10.9	5.2	3.8
1975	10.1	6.7	4.5

In comparison with other large business firms, cooperatives are small. However, the above data do show that cooperatives, in total, in the five selected industries have grown at about the same proportion as the largest comparable noncooperatives over the period 1970–75.[33]

METHODS OF GROWTH. Growth is internal or external. Internal growth includes an adjustment of operations to help accomplish selected strategy through construction of new facilities, increasing membership or business volume, adding services, or market development. External growth includes mergers, acquisition, consolidations, joint ventures, and multi-cooperative arrangements.

In marketing, Philip Kotler breaks up both internal and external growth opportunities for companies into three major classes: intensive growth, integrative growth, and diversification growth.

Intensive growth includes market penetration, market development, and product development. "Intensive growth makes sense for a company if it has not fully exploited the opportunities latent in its present products and markets."[34]

Integrative growth consists of expansion through integration such as backward, forward, and horizontal. Integrative growth would be most applicable where "(a) the basic industry has a strong growth future and/or (b) the company can increase its profitability, efficiency, or control by moving backward, forward, or horizontally within the industry."[35]

Diversification growth, discussed later, is a way for a company to go beyond its present markets.

Cooperatives studied have built up a great deal of growth momentum. A certain part of this growth, however, has come not from any major efforts to achieve growth but as a result of cooperatives being in the right place at the right time. For example, several farm supply cooperatives had already integrated backward into crude oil and fertilizer production before the recent shortage. When the shortage occurred and prices moved up to extremely high levels, the growth rate for these cooperatives in terms of gross sales was tremendous.

Managers in farm supply cooperatives were conscious of the need for integrative growth. The general belief was that farmers must vertically integrate backward. They must own the raw materials for fertilizer and petroleum production if they were going to assure future farmers an adequate supply at a reasonable price.

Only a few cooperatives involved in the marketing of farm products were integratively growth minded. Even in the area of intensive growth, cooperatives appeared to have a great need to become more involved in market and product development.

As for internal versus external growth, most managers expressed a desire to grow by external means if possible. They would prefer to buy a competitor's business with existing brands (in marketing) rather than build from the ground up. This also would give them an already existing market, a set of pretrained workers, and often a set of valuable intangible properties.

Many managers also discussed the possibilities of external growth through merger. However, philosophical differences, personality differences, and ties with farm organizations were considered serious constraints to successful merger. Cooperative managers also expressed some concern about how far the Justice Department and the FTC would permit cooperative mergers and acquisitions.

In the past, many cooperatives have depended more on internal growth than external growth. However, local cooperatives with external methods of growth have achieved a more rapid growth than cooperatives growing only by internal methods.[36]

LIMITATIONS TO GROWTH. Factors affecting growth include management and directors, capital, competition with other cooperatives, member commitment, the external environment as influenced by the consumer and public attitudes, and regulations. Growth in cooperatives depends on the quality and, in some cases, the quantity of these factors, any one of which could limit growth.

Cooperative managers in the core sample saw available capital, the legal environment, and/or member commitment as the major constraints to future growth of their organizations. None of the managers believed growth of their organization would be constrained by management or board members. Difficulty of assuring qualified, dedicated, responsible employees was mentioned by one manager as a possible growth limitation.

Academic personnel were not asked to rank the constraints, but in order of number of responses, the order was board members, member commitment, legal environment, management, and available capital. All except available capital were considered by a majority to be major constraints to growth.

Management and Directors. Edith T. Penrose found that one of the most important factors in growth was "enterprising management." Such management is a necessary but not a sufficient condition to guarantee the favorable growth of the firm.[37]

This study found much variability in management philosophy about growth. Experiences and personalities of managers accounted for much of that.

Many cooperatives rely on promotions to management positions from within. Their philosophy is that they would rather have someone who knows from experience their organization and their cooperative philosophy. Some managers believed that this aided employee morale.

Certain disadvantages may develop from bringing in outside talents; on the other hand, "inbreeding" may limit the organization's growth. Outside managers bring in new ideas, a different philosophy, and, in some cases, the expertise to handle new areas. "If you don't bring in people from the outside who have a different, more open and brighter perspective, you can get into trouble," stated one manager.

Several managers believed that cooperative members and even the "directors themselves aren't very well educated" as to the role and philosophy of cooperatives. Some attempts are being made to better educate directors. A number of organizations such as the Minnesota Association of Cooperatives sponsor an annual workshop for regional cooperative directors. Ten to fifteen cooperatives are represented and various topics are discussed by experts. Other state cooperative associations or other organizations could follow this example. Even local cooperatives could join together in setting up workshops to better educate their directors. Directors who lack sufficient knowledge of cooperative organization and operation may limit cooperative growth.

In a recent survey of 1,833 cooperative directors, Gilbert W. Biggs found 72 percent of the directors believed their cooperatives were about the right size. Only 26 percent of the directors believed their cooperatives were too small.[38]

Capital.[39] Many financial experts suggest that a constraint to growth for cooperatives will be their ability to acquire capital at reasonable cost. The problem stems from several sources. First, concern exists throughout much of the economy about a potential capital shortage. Second, cooperatives have traditionally had fewer alternative sources of funds than noncooperative firms. Third, cooperatives have relied almost solely on one source of funds—the Banks for Cooperatives—for their long-term borrowing needs. Fourth, a lack of financial expertise has existed in cooperatives. Fifth, members generally have been reluctant to invest adequate equity capital in their cooperatives.

A serious capital shortage would have a considerable effect on cooperative growth. Even without a shortage, cooperatives will need a more sophisticated approach to satisfying capital needs.

Managers in personal interviews were confident about their own cooperatives' ability to acquire capital for worthy projects during the next five to ten years. Several based their hopes for future capital on confidence in the Banks for Cooperatives system. Others were confident that alternative capital sources such as commercial banks would be available if the Banks for Cooperatives system failed to meet all needs. One cooperative leader summed up the beliefs of many, "If you are going to grow, you've got to have good earnings, you've got to maintain a good credit rating, and you've got to sustain yourself in a very strong financial posture. If you can do this you'll obtain capital in one way or another."

Questionnaire results yielded opinions different from those expressed during personal interviews. The majority agreed that capital was one of the major constraints to their own organization's future growth and to the future growth of cooperatives as a group. However, in response to another question, the majority agreed that cooperatives in general would not have any more difficulty acquiring capital than would noncooperative firms. Thus they appeared to be aware of the potential capital shortage problems for all American businesses. Managers, however, did not see themselves as having any relative disadvantage when compared with noncooperative businesses.

Academic people were evenly split with respect to their agreement or disagreement on the potential constraint of capital on growth of cooperatives in general. The situation as viewed from a regional cooperative, already financially sophisticated and with a good equity position, may appear quite different from the situation the academician sees. The academician may be more familiar with the problems of small locals than with the regionals. Locals do have more capital constraints than do regionals.

If farmers look to their cooperative system to develop more sophisticated strategies, greater amounts of capital will be required. Many cooperatives depend on the Banks for Cooperatives system for their source of capital. Some of these cooperatives, however, have reached the bank system's lending limits.

Many of the cooperatives nearing the system's lending limits are seeking alternative sources of capital. Others now safely within lending limitations of the bank system should upgrade the level of training of their financial staff and acquaint directors and members with the need to seek alternative sources of credit. Without such precautions and long-range planning, many cooperatives could find that capital, its cost and its availability, will be a severe constraint to growth.

Competition with Other Cooperatives. Another limitation to growth of cooperatives results from their unwillingness to compete for members with cooperatives in adjoining geographic areas, thus representing an orientation to intensive growth. Almost all cooperatives fairly well confined their raw product procurement and sale of supplies to a particular geographic region. Several managers expressed their wish not to expand their membership area into territories of other cooperatives without first discussing the idea with the cooperatives in question. If cooperatives are to continue growing through integration and coordination, greater competition among them will be inevitable. Some cooperatives have deeply penetrated their own geographic membership area and find it necessary to diversify and expand into other geographic membership areas for growth.

Some dissatisfaction was expressed with the expansion of cooperative organization that goes wherever it "darn well pleases." Managers felt strongly about a cooperative that did not respect the implied code of behavior.

Several questions need to be addressed here. One, why should cooperative managers believe their geographic membership area should not be penetrated by other cooperatives unless it is agreed upon? Is it because other cooperatives can offer better and more services which would put great pressure on some managers to compete? Two, would farmers be put at a disadvantage if there were greater competition among cooperatives in certain areas? Would the competition result in a duplication of resources? Three, how could cooperatives work closer together and yet not be constrained by limits to their membership area?

Some cooperative leaders argue that farmers were given the legal sanction to organize cooperatives to lessen the competition among themselves. However, competition among cooperatives is really competition among farmers who are members of the cooperatives. Thus competition among cooperatives goes against the whole philosophy of cooperation.

A number of managers believed that greater competition would exist among cooperatives in the future. Much will depend on selected strategies.

Consumer and Public Attitudes. Consumers and the general public do affect the growth of cooperatives. "The consumer movement," stated one manager, "is taking on greater emphasis than in the past. We may have more to fear than in the past. They [the consumers] are prone to attack the agricultural cooperatives. It's mainly because they don't understand us."

Consumers want stable food prices as well as standards and safeguards. As long as these are forthcoming, consumers have fewer problems with farmers organizing to help themselves. The concern comes when food prices rise and consumers believe it is a result of excessive market power by cooperatives. In 1975, for example, the Consumer Federation of America adopted a resolution that stated: "The immunity from the antitrust laws conferred by the Capper-Volstead Act for agricultural co-ops should be *limited* to the cooperatives which do not possess excessive market power."[40]

Cooperatives do need to improve their acceptance and public relations by identifying and making full use of their commonalities with their "city cousins" and with consumers in general. In the core sample, all but one manager agreed to this need. The response by the academic personnel also favored the need for cooperatives to work closer with consumers in general.

To handle some of the consumer and public relations problems, several cooperatives have hired a person full time to work in this area. A few other cooperatives are giving the idea some thought. Consumerism, however, is not just a passing thought. The consumer relations person must be on the management team and be expected to project consumers' needs and desires into management decisions.

Legal Environment.[41] The vagueness of certain legal issues and the uncertainty of what regulators expect and how they react to specific issues were mentioned by managers as constraints to cooperative growth. As a result, managers were hesitant in some cases to expand certain areas of their cooperatives' operations.

For example, in some commodities, cooperatives have obtained a relatively large market share. By expanding their market share, some cooperatives would give farmers greater bargaining power. The problem, however, is that as cooperatives gain large market shares, their performance comes under close scrutiny by the regulatory agencies. Managers were not concerned with being scrutinized but were concerned with the money and time they would spend if they had to go to court to defend every move they make. Thus some managers were constrained by their concern of obtaining too large a market share.

This raises important questions. How hard should cooperatives push against certain regulations or areas in the legal environment in their growth, and yet continue to maintain their special legal treatment?

Another limitation to cooperative growth is the regulations imposed through recent environmental movements. Some managers believe that environmental movements have had a much greater effect on cooperative growth than consumerism. The environmental issues, however, are a subset of consumerism. Consumers concerned with air pollution and depletion of natural resources have been able to get regulations passed that will help clean up or at least maintain the environment.

New environmental regulations have driven up the costs of operations for most businesses, including cooperatives. The additional expenditures made for control of emissions do take away from investments that could be used for improving or adding new services. The following two examples give some idea of the impact that environmental issues are having on costs.

One manager told the authors that environmental regulations are causing too many cost increases. "The big cost increase came from having to backguard our equipment. The cost of doing this was nothing, $100,000, but the cost of cleaning up every day after every shift and having to take those guards off every time increased our cleanup costs—in some cases double. That cost goes to somebody, either the consumer or the farmer."

Another cooperative organization bought some ground next to a river several years ago in hopes of building a grain elevator on the land. It has been held up on the project for more than three years because the environmentalists were concerned with pollution of the river. An environmental impact statement (the first to be prepared by the organization) had to be prepared since the project involved opening the bank of the river. One of the managers believed that the delay on the project by the environmental movement has cost them "several millions of dollars mainly because we haven't been able to use the facilities."

Another concern of managers was the impact that the Occupational Safety and Health Act (OSHA) may have on their organizations. New regulations and standards for safety and health require increased costs as well as continuous effort on the part of managers to see that their organization continuously meets the set standards and regulations.

Cooperatives have no special legal treatment with regard to environmental, safety, or health regulations. Cooperatives operate with the same constraints as noncooperative firms. But cooperatives, especially marketing cooperatives, are often smaller than comparable noncooperative firms. Thus future growth of the small cooperatives may be seriously limited if the available capital for expansion has to be invested in machinery, equipment, or services to meet environmental, health, or safety standards.

TAXATION. Cooperative taxation has been associated with much controversy, misunderstanding, and confusion. The IRS interpretation and application have added to this. Taxation issues are important to cooperative adjustment strategies because of both positive and negative effects they may have on cooperative growth. Positively, cooperative tax legislation can allow cooperatives a healthier cash flow. Negatively, this legislation requires that cooperatives adhere to a fairly strict set of regulations that affect operation and thus their strategies.[42]

Cooperatives, because of their patron-owner relationship, are allowed certain tax deductions not available to other business corporations. The most important of these deductions, the exclusion of patronage refunds under Subchapter T, is available to both agricultural and nonagricultural cooperatives. The deduction results in a single tax on net margins that is payable either at the patron level or cooperative level.

The tax deductions have a significant impact on providing the equity capital necessary for cooperative growth. Farmers are generally not noted for their willingness to provide equity capital for cooperative growth.[43] The federal tax laws provide a built-in means of doing this while also guaranteeing a direct membership claim on allocated equity capital. Also cooperative tax deductions have a significant effect on capital availability since they help make up for otherwise limited capital sources available to agricultural cooperatives.

Yet few cooperatives would be forced out of business if they suddenly lost their present tax status. Several managers interviewed pointed out that since cooperatives operate on a cost-of-business basis, they could adjust their methods of charging patrons for goods or services to minimize the tax burden if Subchapter T were lost. However, this would cost them their current primary source of their equity capital. Also, many cooperatives have given up their section 521 status, which also permits them to deduct dividends on capital stock and income from nonpatronage business.

Loss of the Subchapter T status would however, seriously hurt the financial position of most agricultural cooperatives. Many leaders believe that loss of the favorable tax status would likely have a very damaging effect on the ability of cooperatives to accumulate capital. Capital availability is especially crucial to cooperatives considering that their size as well as their growth generally lag behind those of ordinary business firms.[44]

Is Growth an Adequate Strategy for Cooperatives?

Growth per se may not be an adequate strategy for cooperatives. If cooperatives are to help the family farmer survive, farmers must select a strategy for their cooperative that is more explicit. Some cooperatives need to do more than simply adjust their operations in line with current business conditions. Some cooperatives, such as the first handler cooperative, may need to integrate into processing and marketing or provide the leadership and resources to maintain or improve an open market to assure the farmer a more equitable price for his product.

Farmers through their cooperatives must select a strategy other than growth that "fits" into a continuously changing environment. They must select a strategy that will help themselves better meet the changing needs of their larger size farms and their increasing sophistication. Growth can then serve as an aid in carrying out this more specific strategy.

4 General Marketing Strategies

MANAGERS AND OTHERS interviewed for this book thought that the choice of strategies lay in three general farm product marketing strategies: (1) pursuing integration and coordination, (2) bargaining collectively, and (3) maintaining and improving the open market. Specific organizational approaches to these strategies as well as government sanctions and techniques necessary to facilitate the designing of a general strategy are discussed in Chapters 5 and 6.

1. Integration and coordination. Fundamentally, this is a vertically integrated strategy expanded with other specific strategies primarily to aid in coordination and market development. More specifically, the objectives often are to obtain additional savings, more control over quality, and greater market power. Implied is a rather complicated, innovative approach to marketing with a heavy emphasis on businesslike tactics, often involving marketing aims backed by advanced research and development programs, sales thrusts, and substantial volumes of business. Diversification and joint venture notions would usually fit here.

2. Bargaining. This is basically the strategy of cooperative market power. It is the policy base for "self-help" programs for maintaining the family farm unit, protected by a strong power-based bargaining organization. Implied here is a set of strategies that call forth organized group action to affect market or countervailing power, usually backed by public sanction and often built on a commodity orientation.

3. Open market. Implied here is that set of strategies rather traditional to cooperatives and heavily dependent on an open or competitive market backed by a good information base and large numbers of buyers and sellers at first-handler levels. Innovation in this strategy is being stimulated by the teleauction and other electronic exchange mechanisms.

All strategies require some form of group action. Such action can be (1) group pressure being brought to bear for legislative changes; (2) formation of a new producer-based organization capable of directly performing a marketing function; or (3) a shift in the basic operating strategy of an existing producer group.

Two or three of these general strategies could exist within a single industry or even a single firm. Sun-Maid Raisin Growers of California, Kingsburg, California, a vertically integrated marketing cooperative, and the Raisin Bargaining Association, Fresno, California, a bargaining group, represent two general strategies in a single industry. Many of the milk cooperatives combine the bargaining and the integration and coordination strategy within one firm.

These three strategies should not necessarily be thought of as a complete set of all possible alternatives but as the key ones that can be brought about by producers through their cooperatives.

Integration and Coordination

CONCEPT AND THEORY. The integration and coordination strategy combines two or more stages of the production-processing-servicing-marketing complex under one management.[1] Integration can be performed horizontally or vertically. But vertical integration is usually the basis for this general strategy.

In horizontal integration, stages at *one marketing level* are combined, usually to increase market power or to achieve operating economies of scale; an example of this is the merger or consolidation of three cotton gins to form a single more efficiently operating unit. Horizontal integration also occurs among a large group of farmers when they market raw products through a cooperative.

In vertical integration, one company combines ownership or control of different stages or levels of production or distribution.[2] Vertical integration may be backward toward the raw material for farm inputs or forward toward the consumer. Some supply cooperatives (Agway is an example) engage in backward integration through ownership of feed and manufacturing plants, phosphate mines, and oil wells. Forward integration can involve processing, transportation, and / or other marketing activities.

A cooperative can integrate through outright ownership of the different stages of the production-marketing process or through ownership of the product itself through the various pricing points. Integration through ownership of stages is the most common approach and can be achieved through external expansion (purchase of existing facilities or merger with another firm) or by internal expansion (growth through construction of new facilities or provision of new services).

Coordination through annual fixed price contracts between farmers and their cooperatives, or between cooperatives and noncooperative firms results in an effect similar to vertical integration. Capital requirements and risks are generally less when compared to integration through ownership. However, integration through the contractual route does not allow for possible efficiencies of single management coordination of two stages.

Feed milling and other facilities used by members of Moroni Feed Company in Utah—a fully integrated turkey producing and marketing cooperative. Capacity of the mill is 600 tons a day.

Neither does fixed price contractual integration allow for any sharing of profits or risks among owners of separate stages, except in the annual allocation of patronage dividends to all members of a cooperative.

Integration can also be achieved through joint ventures. In this case, a contract is made between two cooperatives or a cooperative and a noncooperative firm to provide a specific marketing function.

Theoretically, a cooperative does not have to have commitment from its members to be vertically integrated. But a cooperative could probably not approach this strategy effectively without involving its members directly in its marketing functions. The closeness of this tie can be seen as a fairly true measure of patron support, a potential asset that may present special opportunities for cooperatives to integrate forward into the food product marketing system.

CURRENT SITUATION. Twenty-two percent of United States crop and livestock production was produced under contractual or integrated arrangements in 1970.[3] This was 14 percent for crops and 36 percent for livestock. These figures, especially in livestock, have probably increased since 1970. Indications are that most of this vertical coordination is from the top down. Agribusiness firms in highly concentrated consumer sales markets are integrating backward to acquire control over their supply.

Many cooperatives perform farm product marketing functions. In 1975–76, 4,658 (61.8%) of the 7,535 agricultural cooperatives were primarily marketing farm products.[4] Usually the cooperative purchases the raw product from the open market. It may enhance the competitiveness in the raw product market by acting as another buyer for the product. It may add value to the commodity through processing operations. However, cooperatives seldom achieve true vertical integration very far upward toward the consumer level.

Vertical integration in cooperatives has generally not been carried to the point of retail store ownership, although the idea has been seriously considered. The Ohio Farm Bureau's proposal in the 1960s to purchase the Great Atlantic and Pacific Company (A&P) was an example. Martin A. Abrahamsen has said that "most marketing cooperatives do comparatively little processing, integrate operations to a minor extent, and achieve only limited market penetration."[5]

Some striking exceptions to this general picture exist. The nine cooperatives below illustrate a wide range of commodity areas in which cooperatives use the integration and coordination strategy. Individual examples are in several cases only a portion of a much more complicated overall operation.

American Cotton Growers (ACG), Lubbock, Texas, is integrated in cotton through marketing agreements with growers and ownership of cotton ginning, cotton storage, and cotton spinning facilities. ACG also markets baled cotton lint. Finished cotton cloth is sold to clothing manufacturers, and cotton lint is sold to domestic and foreign textile mills. Sales of ACG brand cotton lint are made in both domestic and foreign markets.

American Crystal Sugar Company, Moorhead, Minnesota, is integrated through marketing agreements with growers and ownership of a sugar refinery. The cooperative sells in bulk and also sells branded consumer packages to retail stores. Sales are in both domestic and foreign markets.

Dairymen, Inc. (DI), Louisville, Kentucky, is integrated in milk through marketing agreements with producers and ownership of milk processing facilities. It also acts as a bargaining association to bargain for milk prices at first sale level. The cooperative sells processed milk in retail stores. DI is a good example of a cooperative integrated both vertically and horizontally.

Land O'Lakes, Inc., Minneapolis, Minnesota, is a large diversified marketing and supply cooperative. It is integrated in turkey operations through marketing agreements with growers, ownership of turkey processing facilities, and sale of branded turkey products to the consumer through nonowned retail outlets, to food service firms, and to institutions.

MFC Services (AAL), Jackson, Mississippi, is integrated in poultry operations through grower marketing agreements. It also contracts to supply the grower with raw materials for poultry production. The farmer supplies fixed facilities and labor for grow-out. MFC operates three poultry processing complexes and sells both branded and private label products to retail stores, the food service trade, and institutional markets. Some poultry products are also sold on the international market.

Riceland Foods, Stuttgart, Arkansas, is integrated in rice operations (as well as soybeans) through grower marketing agreements and ownership of rice storage, drying, and milling facilities. The company sells rice in bulk (export and domestic) and in branded packages for consumer sale.

Tri/Valley Growers, San Francisco, California, is integrated in processed fruits and vegetables through marketing agreements with growers, ownership of can making, fruit and vegetable canning plants, and sale of branded and private label canned goods in both domestic and foreign wholesale markets. A multi-cooperative arrangement (California Valley Exports) with CalCan handles export sales of both organizations.

Sunkist Growers, Inc., Van Nuys, California, is a federated cooperative with marketing agreements between its growers and its local associations. The cooperative markets fresh and processed citrus products in both domestic and export markets. Products are marketed in bulk unidentified form, under the Sunkist label, and under private label.

Other excellent examples of vertically integrated supply-marketing or marketing cooperatives exist. Most of these, except in fruit and vegetables, are marketing the same products as in the examples above. Although all these examples include some form of commitment between grower and cooperative, commitment itself is not a necessary ingredient for vertical integration.

This book places special emphasis on farm product marketing by diversified regional supply-marketing cooperatives. The supply-marketing combination is common. For all local and regional marketing cooperatives, 58 percent handled one or more types of farm supplies in 1973–74. But, of all farm supply cooperatives, only 26 percent marketed farm products.[6]

Vertical integration through federations of locals is often found among supply-marketing cooperatives. The cooperative grain marketing system is a good example. Elements of common ownership exist among cooperatives organized in this manner, but this type of integration is probably not coordinated to the extent necessary to achieve full integration benefits.

Some regional supply-marketing cooperatives have become vertically integrated through direct grower marketing agreements, e.g., MFC, Land O'Lakes, and Riceland. Others such as Farmland Industries (Farmland Foods is the marketing subsidiary) operate processing plants and sell finished consumer products but acquire their farm products through open market purchases from member farmers. Cooperatives such as these have not yet found a way to tie their members directly into their marketing operation.

One reason supply-marketing cooperatives have failed to have much success with vertical integration is that their membership is made up of a diversified set of growers. This wide range of grower interests makes it difficult to start a specialized vertically integrated marketing program because of equity problems from members not producing the commodity.

Supply-marketing cooperatives may also be at a disadvantage in vertical integration because of their frequent heavy orientation to supply sales. Several managers said that their marketing programs were considered by management as a way to sell more feed and other farm supplies. From a management point of view this belief is quite understandable. Sales of supplies often generate higher and more stable net margins than do farm product marketing sales. In fact, some supply-marketing cooperatives have lost money because of the volatile farm product market prices. However, it may be that supply cooperatives have a role to play in the farm product marketing area because of the more predictable margins of the supply business.

Land O'Lakes is a diversified supply-marketing regional that has developed a program that may solve some of these problems. This organization, basically a marketing cooperative to begin with, merged with Felco, a supply cooperative, giving it a substantial interest in the farm supply area as well. Recently, the integrated turkey operations have been combined so that the entire turkey program from selling farm supplies to turkey producers, to processing, to final product sales has been combined into one profit center. This approach is quite new for Land O'Lakes. Nevertheless, several points can be made from this organizational setup: (1) coordination between supply and marketing activities are facilitated instead of allowing undue emphasis on one area at the expense of the other; (2) the turkey division can then reap the full advantages of such coordination; and (3) equity conflicts with other producers who were also served by Land O'Lakes should be minimized because gains or losses should be absorbed entirely by turkey producers that are a part of the turkey division.

Consideration must also be given to problems of containing marketing and supply activities within the same organization. Several levels of coordination of these activities are being practiced. Approaches vary from having separate organizations with a common general manager to having all aspects of the supply and marketing activities combined within one separate entity. Between these extremes are supply organizations that have formed subsidiary operations to carry out the marketing activities. The subsidiaries often have separate membership and management but are still under the overall control of the parent organization and often receive their funding through loans from the parent.[7]

Essentially six methods were observed whereby a supply cooperative could get into the marketing of farm products. They are listed in order of increasing coordination between supply and marketing functions with the exception of the last example.

EXAMPLES OF SUPPLY-MARKETING COORDINATION

Cooperatives	Organizational Structure
FS Services, Inc., and Illinois Grain Corporation	1. Separate regional marketing and supply organizations with a common management only at the top level. Organizations share some overhead functions. About 25 percent of local cooperative members of FS Services are also members of Illinois Grain Corporation.
Agway, Inc., Pro-Fac Cooperative, Inc., and Curtice-Burns, Inc. (a joint venture)	2. Pro-Fac, with its own membership, is a separate organization from Agway, a supply cooperative. Pro-Fac owns and leases plant and equipment to Curtice-Burns, an independent commercial enterprise that is a majority-owned subsidiary of Agway, Inc. Pro-Fac contracts for raw products with member-growers and delivers the products to Curtice-Burns according to established agreements. Agway has control of Curtice-Burns through Agway's majority representation on its board. A representative of Agway also is elected to the board of Pro-Fac.
Indiana Farm Bureau Cooperative Association, Inc. (IFBCA)	3. IFBCA, a supply cooperative, contracts with a grower for raw farm products (turkeys) and farm inputs necessary for production. The cooperative then contracts with a processor to take members' products. The supply cooperative does no processing of its own.
Farmland Industries, Inc., and Farmland Foods, Inc.	4. Farmland Foods, a majority-owned subsidiary of Farmland Industries, markets meat products for its members. Processing is done by Farmland Foods in some cases to consumer level. Farmland Foods has no producer contracts with its members.
Land O'Lakes, Inc.	5. Supply and marketing activities are coordinated within one organization. The cooperative contracts with growers for raw farm products. The cooperative then processes the product and carries it forward in some cases to the consumer level.

The following example is unique enough to be listed separately:

No current example
6. Supply-marketing cooperatives under completely separate management and ownership form a marketing agency in common to jointly market a product under one label. This can be thought of as a joint venture between cooperatives.

The majority of the managers agreed that good rationale exists for coordinating supply and marketing activities within one cooperative. They also felt that cooperatives as a group have failed to gain the savings possible through this kind of coordination.

One possible means of structuring integrated and coordinated marketing organizations is to form a multi-cooperative organization.[8] Several highly successful multi-cooperative organizations exist among supply cooperatives for the purpose of backward integration into the production of plant food. CF Industries, Inc., Long Grove, Illinois, is an example of a multi-cooperative interregional owned by nineteen regional cooperatives for the purpose of fertilizer production. In the farm product marketing area, these joint arrangements are not common. Where they do exist, they are often set up to provide transportation fleets or distribution terminals, not to make independent product sales; Agri Trans Corporation, Long Grove, Illinois, and Ag Foods, Inc., Fleetwood, Pennsylvania, are two examples. A few multi-cooperative organizations do exist in the commodity marketing area, however. Successful examples such as Soy Cot Sales, Inc., Des Plaines, Illinois; Amcot, Bakersfield, California; and Farmers Export Company, Overland Park, Kansas, illustrate possible multi-cooperative organizations in this area.

REQUIREMENTS FOR SUCCESS. Many factors contribute to success in the integration and coordination strategy. Specific requirements for success vary by commodity, but some play an important role regardless of commodity.

Managerial Expertise. Management must have the ability to control effectively what could be considered two types of businesses within one firm. Obviously, this requires more sophisticated managerial skills than if the firm is operating only one stage. Often existing management is capable of performing the additional role, but the increased number of managerial functions will require an increase in personnel at top management levels. In addition, specialists will probably need to be added. Increased responsibilities will usually require higher salaries for top management.

Cooperatives have had difficulty acquiring the management talent necessary for successful vertical integration. Peter Helmberger wrote: "There is a reasonable presumption that farmers have expertise in farming but very little if any in the management of marketing, manufacturing, or retailing enterprises. The dim outline of dilemma becomes quickly apparent. To the extent that farmers participate in the leadership role, they may contribute to poor decisions and hamstring management; to the extent they don't, ownership is separated from management."[9]

The problem may be compounded by an unwillingness to pay salaries competitive with those paid by noncooperative marketing firms. This is compounded by the potential difficulty of working with cooperative board members who have insufficient knowledge of marketing. Also, board members may be unwilling to give management the freedom required to do an effective job. Managerial freedom has its costs, as implied by Helmberger. The researchers encountered several highly successful (in terms of sales and net margins) vertically integrated cooperatives where question arose as to the priority of member interests in contrast to company net margins and return on investment.

The trend seems to be toward a more capable and experienced team of managers in cooperatives. Where it still is not possible to acquire the management skills needed, cooperatives often turn to mergers, acquisitions, or joint ventures to acquire top managers.

Necessary Management Concepts. Success of marketing organizations lies primarily in the capabilities of top management, although location or structural advantages are important. Management needs the necessary experience and must have the support of the membership and board of directors. Management must practice certain key marketing concepts for a successful vertical integration strategy. The most important of these concepts are:

1. Orientation must be toward marketing a product rather than simply selling everything delivered. "Selling focuses on the needs of the seller, marketing on the needs of the buyer. Selling is preoccupied with the seller's need to convert his product into cash, marketing with the idea of satisfying the needs of the customer by means of the product and the whole cluster of things associated with creating, delivering, and finally consuming it."[10] Marketing decisions are based on information derived from the market rather than from the farm.

Questionnaire results indicated that nearly all cooperative managers believed that the strategy of most marketing cooperatives was oriented to selling. In the future, survival of the processing type cooperatives will be highly dependent on the adoption of a marketing strategy rather than a selling strategy.

2. Management must be prepared (and allowed) to coordinate marketing functions under its ownership (or contractual control). Many cooperatives could be considered vertically integrated in structure, but coordination between stages is so loose that many benefits of vertical integration are not achieved.

3. Management should have a comprehensive information system and a long-range planning effort.

4. Management should be prepared to supply customers with a variety

of related products and services. This may require occasional open market purchases and/or nonmember purchases to maintain an even level of service or to fill a product line. It may also require that highly sophisticated services such as computerized information and/or purchasing terminals be provided customers at the wholesale level. Some cooperatives are already moving into this area. Citrus Central, Inc., Orlando, Florida, believes that its goal is to better its services rather than to increase sales. To help achieve this goal the cooperative is setting up a computer system with terminals in buyers' offices to provide information on each buyer's account.

5. Management must be oriented to the need for research and development (R&D). The vertically integrated firm is operating closer to the ultimate consumer. Associated with this is an increased need for product innovation. Firms with a marketing rather than a selling orientation are continually looking for ways to improve their product and thereby satisfy their customers. The cooperative operating among vertically integrated firms cannot ignore this need. This type of firm is also better able to utilize an R&D department since its multi-stage operation provides more opportunities for making use of the results.

Producer Commitment. Since farmers are at the same time owners and users of the cooperative it seems logical that they would willingly commit their products, along with their capital. Yet cooperative leaders typically find farmer commitment very difficult to achieve. Farmers have not considered their cooperative as an extension of their farm business to the extent that they will commit their product ahead and/or make substantial capital commitments. Some notable exceptions exist as were indicated in the examples above. All the cooperatives discussed had a membership willing to sign agreements guaranteeing delivery of their crops to the cooperative. In at least one case, American Crystal Sugar Company, producers also were required to contribute a substantial sum ($100 per acre) for initial equity capital.

Commitment may need to be in the form of a marketing agreement guaranteeing the cooperative that it will receive production on a continual basis.[11] Such an agreement usually does not include a fixed price sale.[12] Instead, price is based on some type of seasonal pool average or formula. The marketing agreement would provide the cooperative a coordinating role over supply similar to the coordination possible through ownership. For those farmers not owning a raw product, commitment might be in the form of an agreement for the farmer to supply fixed facilities and labor necessary to produce a commodity owned and marketed by the cooperative. Marketing agreements enable cooperatives to achieve much of the control benefits of ownership without having to invest directly in farm production.

Product commitment is related to financial commitment. Participation in a pool, for example, often means delayed payment for the farmer but lower seasonal capital requirements for the cooperative. Farmers, expecting the cooperative to build a successful marketing organization, also know they must provide the necessary long-term equity capital. With a financial stake in the cooperative, they are more willing to stick with it and commit their production year after year.

Commitment is also essential between locals and the regional federated cooperative. The regional federation cannot achieve the full advantages of vertical integration without being able to coordinate product flows among the farmer, the local, and the regional cooperative. Few local grain cooperatives have marketing agreements with their regionals or with their farmer members. But one grain federation, Far-Mar-Co., has recently initiated a marketing agreement program that commits the farmer's production to the local and then in turn to the regional.

A committed supply gives management the ability to market products in a more timely fashion, in both domestic and export markets. With a committed supply, more product can be sold in advance of harvest with a higher average price. The cooperative can also enter into long-term commitments to supply other firms owning expensive processing facilities. An assured supply is also essential to guarantee economical utilization of expensive processing facilities and transportation fleets owned by the cooperative.

In exchange for the possibility of higher returns through better cooperative marketing, committed producers would have to give up their freedom to "play the market" in the sense that they are able to choose among various buyers. They might also be asked to live with postponed income while progress payments were being made during a seasonal pool period.

Member Relations. Communication with members is a special problem the noncooperative firm does not have. As the vertically integrated cooperative gets into areas further away from production, its activities become more and more unfamiliar to the farmers. They have only a limited knowledge of what it takes for their cooperative to compete. Yet, as members of the board of directors, they must provide the guidance and support necessary both to protect their interests as farmers and to assure success of the cooperative.

In a 1974 study, 78 percent of the cooperative members polled agreed that there is a risk that a cooperative can get so big that it may not understand the needs of the farmer. Half of the respondents considered it a fact of life that the bigger a cooperative gets, the further removed it is from the farmer.[13]

One manager stated that "cooperatives, as they get bigger, must increase their sensitivity to the membership." Without membership support, the cooperative loses its close ties with its owner users. It thus loses one of its important advantages over noncooperative concerns. An effective member relations program is especially important in the vertically integrated cooperative and cannot be left to chance.

Available Markets. Organizations planning to add a vertical dimension need to consider carefully the nature of the product markets at the newly integrated level. Management must make plans beforehand to maintain, broaden, or possibly segment the market. If no market is readily apparent, the cooperative needs to study the feasibility of developing one before vertically integrating.

Legal-Political-Social Environment. Vertical integration has been said to face less pressure from antitrust regulatory agencies than does horizontal integration. If the industry is dominated by a few large firms, vertical integration may enhance competition.[14] It is also true, though, that forward integration can to some degree reduce competition. The vertically structured firm, depending on its size and share of the market, has the potential ability to foreclose markets, use leverage, discriminate in prices, control production, create barriers to entry, and segment markets.[15] The effect this power can have on the competitive environment must be recognized. Public attitudes and government policy toward vertically integrated structures will continue to have great effect on the success of this strategy for cooperatives.

ASSESSMENT. The *advantages* of backward integration include the reduced risk and savings associated with the assurance of obtaining production supplies of the required kind, quality, amounts, and at the required time.[16] Savings are also often possible from the firm's ability to produce its supply requirements more cheaply internally than by acquiring them from the open market. These factors have probably been the motivation behind the backward integration of the supply cooperatives into production of fertilizer.

The motivation for forward integration toward the consumer is tied to the advantages of coordinating under one management raw product procurement, processing, and sale of consumer products. The cooperative having control over these forward stages has increased the certainty of its market outlet and its control over its merchandising channels. It has added value to the product and has probably enhanced the product's profitability. The firm has brought itself closer to the consumer where its advertising, promotion, and market influence will have greater impact. A further incentive for forward integration is the opportunity to reduce costs through economies of scale or through elimination of interstage transfer costs. Returns are often more stable when generated from forward markets. Forward integration also improves the setting for development and use of new technology or development and promotion of new consumer products.

Cooperatives may find the greatest potential benefit of vertical integration to be that the farmer is brought closer to the consumer and the final demand for his product. Marketing cooperatives are then better able to relay important demand signals back to the farmer or the production side. Both farmer and cooperative then become more market oriented.

Vertical integration also carries with it some possible producer *disadvantages.* In instances where farmers have contracted their crops with a vertically integrated cooperative they would not be able to sell on the open market. A compensating factor would be their risk reduction from market price fluctuations, especially if their cooperative operated a seasonal pool.

At least one cooperative, Calcot, Ltd., Bakersfield, California, has eased the problem by allowing the use of a "call pool." The farmer still commits his production, but he can play the market by fixing the price at a time of his own choosing—usually within a predetermined time interval.

Possibly the most serious disadvantage with vertical integration is the large investment required. The farmer is required to contribute capital to finance and maintain the vertically integrated system. However, some capital requirements might be satisfied by taking equity in merging or consolidating cooperatives and forming a base capital pool. Other options might initially require retention of a large amount of patronage refunds, per unit retains, or direct assessments based on historical delivery patterns. Cooperatives may need to seek new sources of debt financing and use new and complicated approaches to acquire the necessary debt capital for vertical integration. (For more detail see Chapter 6, Facilitating Strategies, "Expanding Financial Sources.")

Members may be subject to much greater risks if a cooperative engages in a substantial amount of contract production. This can become a serious problem if the cooperative has a diversified membership.

A final disadvantage is that pricing points—points at which the transfer of product is made between stages—are eliminated under vertical integration. Groups interested in knowing these interstage transfer prices, such as producers, consumers, and competitors, are then without such indicators of the raw product value. When product prices cannot be reflected at each stage level, producers may have difficulty in monitoring or controlling the integrated operations.[17] Unsatisfactory price information in an industry may also make it difficult to establish proper administered prices in government-sanctioned programs.

It is quite difficult to assess the effect of the vertically integrated firm on society in general. The increased efficiency of the vertically integrated system may result in lower food costs to consumers. But it is also possible that the cooperative could use its strength to restrict supplies marketed or to restrict competition in its markets. However, producers put in closer contact with the consumer should be able to adjust their products to better satisfy the consumer's wants.

Vertical integration may be less susceptible to antitrust persecution than horizontal integration.[18] There have been few instances where vertical acquisitions have been challenged in the food industries. There is, however, legal support for preferring that vertical integration occur through internal growth rather than through external growth. The enforcement of antitrust legislation has clearly indicated that integration achieved by construction of new plant capacity or distribution outlets is the more acceptable approach.[19] The belief is that growth by combination as contrasted to internal expansion is more likely to stifle competition.

Cooperatives, regardless of growth strategy, should be prepared to accept some erosion of public support as they grow larger and more visible. As they integrate forward and become more competitive with other firms, they run the risk of damaging their relations with noncooperative firms.

Growth through vertical expansion, may, however, be more acceptable or may present less visibility to the public than growth through horizontal means.

There should be incentive for vertical integration only so long as the ownership of another stage or product beyond the present pricing point is profitable, either for the added function itself or for the overall organization. Integration may also be accomplished to protect a market even though profitability is not immediately available. The opportunity cost of not vertically integrating may be considerable. Sometimes, however, little logic exists in spending large sums of money only to carry integration to a level where competition is tougher than at the present level.

One cooperative manager expressed this when considering getting into processing: "Processing really is just one more step in the integration chain. We're still up against the stiff competition of retail buyers or wholesale buyers., We've just moved up one more level, and we certainly aren't going to buy Kroger." Most managers did not believe cooperatives needed to go as far as ownership at the retail level.

In the long run, control over marketing channels is the only way to ensure that the farmer's interests are protected in the consumer marketing area. This control can come in various ways. Outright ownership of marketing stages is only one approach. Cooperative managers should also consider the advantages of vertical integration by owning the product through various stages without actual ownership of stages or facilities themselves.

One recent proposal for cooperative integration in the livestock industry suggests that the livestock cooperative perform a coordinating role for the farmer in the product's movement through various production-marketing stages.[20] The system would require commitment of the farmer's product to the cooperative but would allow the producer to maintain his ownership through up to four of the successive stages of the cattle production-marketing complex. The cooperative would have the option of acquiring its own facilities to perform the various marketing functions or it could contract for these services. In either case the cooperative would have a committed product supply and would be in a position to improve the farmer's marketing strength. Producers through their cooperative could improve supply and demand coordination and increase the level of market intelligence.

This type of coordination could be initially performed on a contractual basis for the various processing steps. It could serve as a stepping stone for future ownership control of these stages if the cooperative should determine that such control would yield net margins and put the farmer in a stronger marketing position.

Vertical integration is most likely to be initiated in situations where available marketing or supply channels are being closed to the farmer or where pricing mechanisms are deteriorating. Where marketing or supply alternatives are many, the farmer will usually have little inclination to support a new vertically integrated organization. Cooperatives that wait until market conditions deteriorate before changing their functions are reactors to their environment. They often pick up what a noncooperative has decided is an unprofitable business. Cooperatives justify the move because of the interests of their farmers. This can be a mistake. An unprofitable commercial processing plant will probably be an unprofitable cooperative processing plant. Farmers need to be actors more than reactors. They should anticipate changing market characteristics and be at the forefront with an improved integrated and coordinated marketing system if that is their general strategy.

Bargaining

Bargaining is a broad concept with many facets. It is basic to individual firm management. It has been put forth as a key feature of farm price and income programs. Such use of bargaining raises public policy, legal, and equity issues.

No attempt is made here to treat the whole subject of bargaining; the purpose is more modest and mainly to present bargaining as an alternative strategy that producers can use to help protect their economic welfare through group action. Essentially, the emphasis will be on this strategy as it fits an individual cooperative. Recognized is the fact that expanded federal legal sanctions for bargaining, especially highlighting agricultural cooperatives as the bargaining agents, would go far in welding together massive countervailing power for agriculture in total.[21] Obviously, cooperatives would have an expanded use of this strategy if it were given thrust by agriculture in general. This latter issue is illustrated by expansion of European cooperative bargaining and by proposals seriously considered in the United States in the 1960s and to some extent in more recent bills such as the so-called Sisk Bill.[22]

CONCEPT AND THEORY. Bargaining in agriculture is defined in many different ways. Some managers who contract for a supply of product with local producers say they are bargaining when they negotiate price and other terms of trade with processors. These managers are performing a bargaining function. They represent an association of producers, and their major concern is to provide their member-owners an equitable price and terms of trade for their commodity. Managers will usually negotiate one price and terms of trade for the total crop produced within a given season.

Livestock marketing cooperatives selling on central markets also claim they are performing a bargaining function. Every time their sales representatives meet buyers, they believe they are bargaining. Actually this function provides a competitive atmosphere and is more nearly an example of the open market strategy as defined in this report.

Collective bargaining is usually thought of as a process associated with labor unions. Collective bargaining has been defined as "the process by which wages, hours, rules, and working conditions are negotiated and agreed upon by a union with an employer for all the employees collectively whom it represents."[23]

Collective bargaining, however, has become prevalent in agriculture in recent decades. In agriculture, collective bargaining is a process whereby an agency of an association of voluntary producers representing a particular raw commodity negotiates in an orderly fashion with handlers or processors to arrive at prices and other terms of trade that are in the best interest of those whom the agency represents.

Experiences with collective bargaining have shown that voluntary participation by producers has not been effective. To overcome this, attempts were made by bargaining leaders to establish methods, rules, and regulations giving collective bargaining associations the right to represent all producers in a particular area as their agent, i.e., exclusive agency bargaining.

> The basic concept of exclusive agency bargaining includes as a minimum a bargaining unit consisting of a group of farmers producing a common product; a bargaining association which has the authority to represent in trade all farmers in a bargaining unit, whether members of the association or not; and a set of rules establishing rights and obligations of members, nonmembers and handlers or processors.

> Exclusive agency bargaining differs from voluntary collective bargaining by the fact that the exclusive agency bargains for price and other terms of trade which apply to all who are defined within an association, members and nonmembers alike.[24]

The concern in this book is with the concept of bargaining as a general strategy for agricultural producers. A successful strategy may require exclusive agency bargaining rather than voluntary bargaining. Views of various people involved with bargaining are divided on which approach best satisfies farmers, handlers, processors, and the general public.

CURRENT SITUATION. Currently, bargaining associations are active in such commodities as dairy, fruits, and vegetables. Approximately 230 active producer associations are involved in bargaining, and about 100 of these represent dairy producers.

Many bargaining associations bargain for only a single commodity. Other organizations such as the American Agricultural Marketing Association (AAMA), Chicago, Illinois, are involved in bargaining on a multi-commodity basis. AAMA is a national coordinating and information agency set up by the American Farm Bureau Federation to help state affiliates carry out marketing programs—bargaining included.[25]

The range of opinions on the role of bargaining as expressed by the managers in interviews was wide and varied. The following discussion relates some of their comments.

Farmers do need some type of organization or method of operation for countervailing power. A bargaining association can fulfill this need.

"Bargaining is a mechanism whereby you can get what you're entitled to," stated a former manager of a bargaining association. "A farmer can't afford a loss as easily as in the past. He needs a stable market—good price." The role of bargaining is to "establish the value of raw material—especially crucial in our times of difficult pricing discovery.

"I feel the bargaining idea in agriculture has a big future because I am convinced that contract production is going to come into agriculture quickly. I think in twenty years we will see a big trend in agriculture toward bargaining. In livestock this will happen if we get more contracting and if we lose the price-making system that now exists.

"Bargaining does not really take place until we get contractual vertical integration. . . . in the end we will either have bargaining associations and farmers operating as business people or we will go the broiler route where farmers are wage earners with possibly some physical input of plant but [the farmers] will not be risk takers.

"Where there is need for a bargaining association is where there isn't a dominant cooperative, where a commercial operation doesn't allow growers to share in the proceeds, then there is a place for cooperation among grower-members of processing and bargaining associations."

A former manager of a bargaining association believed that bargaining would be a major factor for agricultural producers in the future because:

    1. Increased concentration in the market must be offset by producer bargaining,
    2. Maintenance of the independent family farm is dependent on some control of the market,
    3. A continuing and increasing need exists for information regarding market forces,
    4. An increasing need exists for producers to remain competitive with other sources of raw material (example—cotton vs. synthetics).

"In the tomato processing industry," stated a bargaining association manager, "the number of buyers continues to decrease either through acquisition or merger. Fifteen years ago there were 54 buyers; we now have 30 including three cooperative processors. Multi-national and conglomerate corporations are prominent in the industry whose main interest may not be in the food or tomato business . . . consequently the decision-making mechanism of these companies may be located as much as 3,000 miles removed from the actual production site. In order for our member growers to receive an acceptable return they must have equal muscle in developing a price that will return a profit to the average grower, rewarding the efficient grower and eliminating the inefficient grower."

Several managers of operating cooperatives thought that the only way farmers should think about bargaining is by farmers owning their own operation and being in the "mainstream of things." The general feeling was that farmers would be better off if they went beyond bargaining and got into processing.

In California, managers of a marketing and canning cooperative expressed the need for bargaining cooperatives to exist alongside the vertically integrated and coordinated cooperatives. In this way the operating cooperatives would not have to become involved in price establishment. They could accept the price established by the bargaining association in negotiations with noncooperative handlers and processors. The cooperative processors themselves would not need to enter into any of the negotiations. In this way, marketing cooperatives would avoid litigation that might occur from suits involving undue price enhancement.

Several academic people expressed views about bargaining as a major strategy. Some of those ideas follow.

"Bargaining will be nearly impossible to achieve in the major agricultural lines such as the grains, livestock, etc. The interests involved are extremely diverse and essentially difficult to organize and coordinate. The commodity groups in which bargaining has been reasonably successful have been those operating in a limited area and with a relatively harmonious member group."

"For some important commodities, there will be little opportunity for large-scale operating co-ops; even if there are, there is need [by a bargaining agency] to represent growers [who are] not members of marketing co-ops in price and contract equity issues."

One professor gave five reasons why bargaining will be a major factor for farmers in the future:

1. Will be the vehicle of price discovery as open or competitive market prices will disappear.
2. Fewer and larger farms.
3. Demands of financial institutions for reasonable return on investment (ROI) per farm firm financed.
4. Improved management skills of farmers.
5. Continued concentration in food industry.

The status of bargaining will depend primarily on the extent to which "farmers" and "ranchers" will own what they produce. If the ownership of crops and livestock by processors, packers, feeders, integrators, chain stores, etc., increases, then the need for bargaining will increase. If ownership by these decreases, bargaining will be a lesser factor.

The following statement from a bargaining association manager helps portray the importance of agricultural bargaining: "Independent farmers are the most efficient producers in comparison to large corporate farming. It is for this reason that bargaining will be a major factor in establishing reasonable prices for agricultural producers who, without a strong bargaining organization, would be at the mercy of the dictates from large corporations and would suffer from extremes in raw product prices.

"Farm bargaining must respond with contract terms, quality of product, and the welfare of the industry it is involved with. A farm bargaining organization which only considers the short term will not survive the long pull.

"Farm bargaining must evaluate many factors with credibility being maintained in order to be successful."

REQUIREMENTS FOR SUCCESS. Bargaining success may rest heavily on one or two factors such as localized production areas or tight producer commitments, but often it will rest on the combined effects of several factors. The weight each plays on success will vary with the situation.

Important factors for effective bargaining are: (1) enabling legislation, (2) member commitment, (3) production control, (4) buyer recognition, (5) leadership, and (6) market information.

Legislation. Many people consider legislation the most important factor affecting future success. They would like to see legislation, either federal or state, that would give bargaining associations exclusive agency bargaining rights for all members in a designated area. Buyers would then be forced to negotiate with bargaining associations in good faith. Laws would spell out certain procedures for arbitration if an agreement between the association and buyers over price and terms of trade could not be reached.

Member Commitment. At the present time, some bargaining associations lack member commitment and control over supply. Buyers do not recognize the association because they are able to buy raw products from nonmembers.

Effective bargaining with voluntary membership requires grower confidence that the bargaining association is operating in their best interests and that the price is fair and equitable in terms of the long-run success of the industry. Lack of grower commitment is deadly to bargaining strength.

Production Control. "Production control means the ability to avoid the growth of surpluses. This is a long-run supply concept. Effective bargaining usually means higher prices. Higher prices bring on more production. More production dissipates the bargained gains. Control of production is a difficult component in any agricultural policy and bargaining power approaches basically fail to avoid it."[26]

Supply balancing may require growers to limit production to an acreage or volume specified by the association. Marketing agreements would be helpful. The agreements assure the grower a home for his product, and the association is assured the volume of product it can commit to buyers. Good supply control will provide buyers with the necessary quantity of product at a given time and a reasonable price. This will hinder buyers' efforts for finding or developing substitute supply sources.

Buyer Recognition. "Recognition means that the bargaining group must be able to represent the producers at hand and be so recognized by its adversary.''[27] Recognition would probably be best achieved through some type of statutory obligation such as the right for exclusive agency bargaining. Buyers would by law be required to bargain in good faith.

Leadership. Managers of the bargaining association must be aware of both growers' and buyers' needs. They must be knowledgeable and experienced in the art of negotiation and compromise. Association managers must continuously strive to improve relations with growers and buyers. Information on crop conditions and other factors should be made available to members and buyers on a regular basis.

The association should take title to the product. Total responsibility by the association requires managers to do a better job of negotiating and finding a home for the product.

Information. The association and buyers must share some information. In this manner, cost of production, handling, and processing will all be prior knowledge to negotiations over price and terms of trade. With both parties sharing the same information, neither is at an advantage or disadvantage; more rational economic decisions can be made.

Successful bargaining may not be achievement of a satisfactory price and terms of trade in any one year. Success in bargaining has to be measured over the long run, and consideration has to be given to how equitable the whole process is to buyers and to consumers in general.

ASSESSMENT. Most managers, with the exception of those in specialized bargaining associations, said they had not thought much about bargaining and did not expect it to be a major strategy in the near future. Also, policymakers are not pushing it as a major thrust.

Yet bargaining is vitally important to selected producers and specialized producer groups. Also, bargaining in its various forms is providing several functions, many of which are important to all cooperatives and the total marketing structure. Bargaining usually becomes a more significant strategy issue in times more adverse than those farmers have recently experienced. Trends toward increased market concentration suggest bargaining may become more of an issue. For these reasons an assessment of bargaining as a general cooperative strategy is in order.

Six *advantages* are claimed for the bargaining strategy.

First, bargaining is a rather traditional way for farmers to countervail middleman power so as to guarantee a larger share of the margin between farmer and consumer. Often this may be the only expedient, short-run strategy. The bargaining association may be necessary as a substitute if a processing cooperative or a general farm organization is not available to do the job. The specialized or limited nature of the production or marketing of the product may not make it attractive to larger organizations. Thus the bargaining cooperative may be about the only strategy that farmers can employ. Also, it may allow them to get a strategy with limited producer equity. Operating capital is usually acquired merely by assessment of a per unit bargaining charge.

Second, the bargaining association may be the most feasible means to guarantee a market outlet. Some of the reasons given above make this true, but also the bargaining unit may be able effectively to coordinate the supplies. Such a service can attract buyers and yield some savings to the marketing system with additional returns to producers. Cooperatives can easily overstate the savings in assembly and other coordinating functions. However, cooperatives are usually in the best position to provide these functions.

Third, farmers may need to resort to bargaining to assure what they consider an equitable consumer price. Producers have an interest in consumer prices since their farm prices, especially in the long run, depend on consumer reaction. One source of possible long-run bargaining gain for farmers is through higher consumer prices. Farmers may find that if they bargain effectively they can force those with whom they bargain to negotiate more aggressively with consumers. For various reasons, farmers may not be able to build a marketing system to the consumer that is completely under their own control. In such cases, the bargaining strategy may be the most effective way to extract a larger consumer price.

Fourth, bargaining may be necessary to assure a functioning price-making system. Many factors, especially vertical integration, are eliminating price-making points in the market process. Price information dries up and the rest of the marketing system suffers. Public policymakers, including cooperative leadership, may need to consider this issue from the viewpoint of the overall marketing system.

A much more direct issue comes up with regard to this advantage, and it involves interactions among cooperatives. Bargaining cooperatives may well need to coexist with processing cooperatives. A good example of this approach was found in California among three grower-owned fruit and vegetable processors and canners that process both tomatoes and peaches. But the peach and tomato growers also have bargaining associations to establish price and other terms of trade with canners.

Two of the cooperative canners did not enter into price negotiations with the bargaining associations. The cooperatives accepted whatever price the bargaining associations established with the independent canners. They felt that this is "the proper sphere in which price ought to be determined."

The cooperative canners gave several reasons why they had not entered into the price negotiations:

1. They sought two opposing objectives; they wanted to see the grower get a good price for his raw product, but as a canner they also wanted to obtain raw product at a sufficiently low price to show reasonable net margins for the business.

2. They avoided any pressure from the Justice Department and other government agencies in the area of undue price enhancement.

3. Independent canners could not accuse them of paying an initial

price for the raw product that included part of the cooperative canners' earnings.

Fifth, bargaining may be an effective strategy to guarantee supplies to cooperatives with facilities of their own. Many cooperatives have processing or other facilities that must be kept at operating levels necessary to give efficient marketing. They may not want to go fully to the integration and coordination strategy but merely use bargaining to operate one or two plants or to dispose of troublesome seasonal surpluses. Adequate supply may be a result of this bargaining function. These supplies could be used within one cooperative—as an assured source of supplies and monitoring of the market, or as a coordinated program among one or more bargaining cooperatives and one or more operating cooperatives.

Sixth, the bargaining strategy may be a way to work toward either of the other two major strategies discussed in this book.

The effective "open market" of the future may well depend on bargaining units controlled by farmers guaranteeing adequate supply for trading in a competitive framework. Price-making forces can then operate, and price information can be available not only for vertically integrated units but for small cooperatives or individual farmers selling through other marketing channels. Many of these channels have historically had within them central markets to establish price. Other institutions used this information. In much the same way, bargaining associations may be able to provide this service and help maintain other aspects of the strategic concept of open competitive markets.

Movements from the bargaining strategy to the integration and coordination strategy are probably more likely. The bargaining approach could be a stepping stone for growers to get into processing and further marketing. Dairy cooperatives are a good example of bargaining associations that have moved into processing. American Crystal is also another example of sugarbeet growers who extended themselves into processing from a bargaining association.

As an approach to agricultural marketing, collective bargaining may be a very effective short-run solution for farmers to obtain the prices necessary to keep them in business or to realize a reasonable return on their investment. It may also present the first opportunity for producers to get together, form an organization, develop leaders, and begin to work toward a common goal. It may, in fact, be a crucial step toward the future development of an organization capable of aggressively marketing the farmer's product. It has been, and may continue to be, a most logical first step toward the integration and coordination strategy.

Five *disadvantages* are claimed for the bargaining strategy.

First, bargaining simply may not be feasible as a general strategy for many individual cooperatives nor for the cooperative movement overall. The widely dispersed farm enterprises such as most livestock, many grain crops, and some fibers may not be manageable in this strategy. The necessary supply control and producer commitment are too large a job for cooperatives alone. The level of government control and sanction necessary to aid cooperatives in bargaining may be unacceptable to farmers and the general public. Bargaining success leads to its own downfall unless production is controlled in response to the higher prices. Supply control is difficult for many reasons—legal problems and nonmember problems are two formidable ones.

Most managers said that producer commitment was the single most important requirement for bargaining success. Yet commitment waxes and wanes—high in adverse times and low in good times.

Administrative costs and problems of widespread bargaining programs would be difficult to handle. Who would decide the regional conflicts? The logistics of price establishment in a commodity like hogs or wheat raise a problem of the feasibility of the bargaining strategy.

Second, bargaining needs a backup system of support in processing plants, marketing outlets, surplus disposal plans, and such. The question then becomes, Why not go all the way and use the integration and coordination strategy? Why stop halfway? Also, a danger exists that the backup will become the tail wagging the dog. Such backup operations are businesses in their own right. They must be well run. Danger of overextension is a potential problem.

Third, bargaining success is difficult to document. Farm prices may be high enough to give good incomes, but the nagging question to farmers is, What would prices have been if they had not had the bargaining association? In many ways bargaining is a defensive role and results are difficult to establish. Many see the bargaining act as increasing prices but adding no tangible value to the product.

Fourth, substantially strengthened bargaining could face legal constraints. Collective bargaining is sanctioned in current enabling legislation such as the Capper-Volstead Act and the Agricultural Fair Practices Act. But bargaining means price enhancement. The question, then, is, How much enhancement? The meaning of "undue enhancement" under the law must always be in mind. That part of the law has not been well tested, thus the future of bargaining sanctions is not clear. Uncertainty exists about the passage of any federal legislation giving associations exclusive bargaining rights and forcing buyers to negotiate in good faith.

Many people are watching the outcome of the Michigan Agricultural Marketing and Bargaining Act, signed into law in 1973. The act originally was to expire on September 1, 1976. However, even with eight major lawsuits in process, the act was amended on June 17, 1976, eliminating the law's expiration date.

The Michigan law is a benchmark for bargaining legislation. Major provisions of the act allow (1) agency bargaining, (2) appointment by the governor of a five-member Agricultural Marketing and Bargaining Board to administer the act, (3) accredited associations and bargaining with processors in good faith, and (4) compulsory arbitration.

The continuance of the act depends on its constitutionality.

According to Gene Ingalsbe significant allegations include:

—that the act violates the State constitution by exceeding the State's police power;
—that the act is contrary to the guarantees of due process of law in permitting legislative power to be conferred on private individuals;
—that the act is inconsistent with the federal Agricultural Fair Practices Act and violates the U.S. constitution;
—that accredited associations do not meet the requirements of the Capper-Volstead Act and their actions are in violation of the Sherman Antitrust Act; and
—that the Agricultural Marketing and Bargaining Board did not comply with the Administrative Procedures Act in bargaining unit and accreditation procedures.[28]

If Michigan can work out the problems associated with the act, other producers in other states may seek similar legislation. If Michigan cannot find a workable solution to these problems, then other attempts at bargaining legislation will certainly be watered down.

Bargaining associations need to take title to the product not necessarily because of the legal implications but because taking title gives much greater responsibility to the association. The association, explained one manager, becomes closer to the actual functions of the marketplace, and once it owns the product, it has to market quantity and quality. The association, by taking title, assumes a much greater responsibility to work with processors to market a good product at a fair price.

Another factor in the bargaining approach is the move toward use of multiple-year contracts. However, the legality of long-term contracts may become a matter of concern. One organization reported that 65 percent of their contracts with growers were for more than one year. The bulk of the agreements were for two years. Some of the agreements, however, were for eight years. This organization was also interested in long-term pricing with canners. The manager believed that "it adds to stability, strength, and planning." So far, he has not been able to convince the canners of this approach.

Fifth, any large-scale bargaining effort involving several cooperatives raises a question of who will do the bargaining. The whole set of "who-is-to-bargain" questions arise—large versus small, local versus regional, bargaining versus processing, and general farm organizations versus commodity cooperatives. Of course, the buyers also have great concerns about knowing with whom they must bargain. Conflicts between producers of the same commodity add even more concern about the designated bargaining agency.

The bargaining strategy must be kept under study by farmers and cooperative leaders. Economic and political climates affect the interest in such an approach. Some believe that general farm policies could return to more highly regulated forms because of future developments such as recently declining farmer income, need to have greater direct producer impact into foreign trade, intolerable price and income variability, more public subsidy of food production for diplomatic or welfare reasons, protection of the family farm, and a host of others. Past history suggests that policymakers usually give the bargaining strategy consideration at such times. Any such policy use of the strategy puts cooperatives squarely in the middle of making the strategy work.

Cooperatives have varying degrees of bargaining strategy inherent in their very being. This issue becomes more relevant and complicated as structure becomes more concentrated, both in the cooperative and noncooperative sectors. Many issues bring bargaining forth as a cooperative strategy. Adapting to open market deficiencies, expanding market needs for joint selling arrangements, price establishment techniques, supply coordination, and coordination of general farm organization functions with commodity functions are issues that a bargaining strategy may well address.

Thus cooperatives must keep this general strategy in mind at three levels: What is its relevance for each cooperative; what are the issues here for joint strategy and planning among cooperatives; and what may the policymakers do in this area that could involve cooperatives?

If bargaining is to gain in significance for farmers, managers of operating cooperatives will need to work closer with bargaining groups and give them legislative support. It may not be necessary for operating cooperatives themselves to get into bargaining, but they must realize its importance and be willing to work with farmers to assess the general bargaining strategy.

Open Market

CONCEPT AND THEORY. The open market strategy provides an environment allowing cooperatives to provide marketing functions in a competitive system, with very little influence over price, and usually at the first-handler level.

The basic objective of this strategy is to do whatever is necessary to guarantee the operation of the open market pricing system within a competitive market environment. The open market strategy is probably the most difficult to visualize because it includes such a wide range of approaches. One way of visualizing the concept is to consider a pricing point itself—the point at which a product is transferred from one owner to another. Several different points may exist as the product moves from one stage to another in the marketing channel. Each point at which the product changes ownership is a pricing point. The open market strategy does *not* include the possibility of bypassing that pricing point by integration of the two marketing stages. This strategy *does* include the possibility of changing or improving the existing method of pricing. It also includes altering the structure of the industry on the *buyer's* side of the pricing point. The

bargaining strategy, defined earlier, is the option of making structure adjustments on the *seller's* side by forming a bargaining association.

Approaches to the open market strategy all have one thing in common—they are all aimed at maintaining the pricing point on an open market basis. Action would be taken to keep the market alive with a level of trading heavy enough to reflect accurately the market value of the product moving between two marketing stages.

The concept of the open market seems to be only loosely defined. From a producer's standpoint, an open market is a marketing system accessible and available to producers at all times. It offers potential for sales to many prospective buyers. But this theoretical ideal is seldom achievable in the real world. Jack H. Armstrong has proposed a more practical definition: "An open market is a market that provides a producer the assurance when he begins production, without prior commitment, that there will be a buyer for his product when he is ready to market—hopefully at a competitively established price."[29] Local grain markets are an example of open markets. They are accessible to grain producers who can sell their product at any time to any one of several local elevators.

Farmers usually expect an open market system to coexist with a more closed market system such as contract production. This is the case whether

A new method of marketing feeder pigs called Tel-O-Sale is being used by Producers Livestock Association in Columbus, Ohio. Using a conference telephone call, the man on the right auctions off the feeder pigs while the man on the left records the progress.

they typically sell to their cooperative or to private firms. The relatively closed systems depend on active open markets as essential sources of pricing information. But, when given the opportunity, the farmer will often use the more closed system if it seems to offer an advantage.

CURRENT SITUATION. The trend in the marketing of many commodities is the gradual disappearance of the open market system. Commodities most affected include broilers, turkeys, eggs, fruits, and vegetables. Even central livestock markets are in some cases being replaced by contracting and direct buying.

As a substitute for the open market, product is now often contracted directly to processors at planting time. Alternatively, the marketing process becomes integrated through single ownership of at least the production and initial marketing stages. In some areas, large agribusiness firms have integrated backward into the actual production of agricultural products. Farmers, when they sometimes sell their product directly, are also bypassing the central market. In some industries, farmers who are not contracted to a buyer at the time of planting will have no place to sell their crops at harvest.

Remaining open markets are characterized by low volume, lack of competition among bidders, inadequate information, inaccessibility to traders, and a high potential for price manipulation.[30] The effectiveness of the price discovery process is being reduced, hence reducing the competitive conditions for accurately reflecting the forces of supply and demand.

Many administered pricing systems, such as a federal order pricing system of milk, depend on price information or quotations. When the market fails to generate such price information, these programs cannot operate effectively.

The pricing problem is severe where pricing decisions in closed marketing channels are based on product movements in thinly traded open markets. The problem exists whether the pricing is governmentally administered or privately operated based on price information generated from another market exchange. To the extent these thinly traded markets are relied on, the competitiveness of the entire farm marketing system is being reduced.

Forecasts indicate no lessening of these trends. James V. Rhodes has said of the hog industry, "There seems to be no stopping of the trend toward a withering away of terminal markets."[31] Olan D. Forker believes of the future that "an increase in decentralization, in private, closed transaction systems, and in contract commitments of various kinds to trade in advance of price determination will make it more difficult to know whether the pricing environment is competitive or not, and will make it possible for imbalances of power to be exploitive."[32] If these trends persist, farmer access to open markets will rapidly be a thing of the past in some commodities and will be in a precarious situation in others.

REQUIREMENTS FOR SUCCESS. Approaches to maintaining the open market pricing system involve a number of possible activities, including enhancement of the flow of relevant market information to market participants, improvements in grading standards, regulations requiring buyers to utilize open market pricing systems, and the operation of electronic exchange mechanisms.[33] A combination of these approaches may lead to the most effective open markets.

The electronic exchange mechanism will be stressed first because it can contain elements of all these approaches. It works best if certain requirements are met. First, the product to be traded should be homogeneous. Any quality differentials should be uniform enough to allow sorting into standard grades. Second, the product should not deteriorate or change its quality over the trading period. Third, a large quantity of the product must be available for trading on a frequent basis. Fourth, traders must have an interest in a more competitive market system.[34]

Probably the most popular approach for an electronic exchange mechanism is the telephone auction or teleauction system. In this approach pricing is done through the auction method. One of the country's oldest teleauctions is the Dublin Tel-O-Auction in Dublin, Virginia, established by the Eastern Lamb Producers Cooperative.The auction has been used for marketing feeder pigs, butcher hogs, slaughter cattle, feeder cattle, and market lambs.

Previous to the startup of the cattle auction, producers had only a very limited number of buyers operating in their area. The farmers organized an auction using a conference telephone call combined with a state-financed and state-operated system of grades and standards. Buyers are now able to bid on the area's livestock from several hundred miles away in simultaneous two-way communications between auctioneer and bidder. It is no longer necessary actually to visit the area to view the livestock. Animals are moved directly from the farm to the buyer's packing plant. The buyer saves himself the cost of traveling to the area, and the seller saves the time and cost of assembling his livestock at the central market. Producers find the competition to be increased. Experience with the lamb teleauction has demonstrated a 6–10 percent price advantage for sellers due primarily to more competitive bidding.[35]

Current interest in the teleauction approach in the United States has been confined primarily to market lambs and feeder cattle. In all cases, these markets are being used on a voluntary basis. Nationally, their share with respect to the total market has been quite limited. However, where they exist locally, they represent an important market alternative.

A much larger auction is the teletype auction operated by the Ontario Pork Producer's Marketing Board, headquartered in Toronto, Canada. All butcher hogs within the province of Ontario, Canada, must, by provincial law, be sold on this auction. This setup was established by producer referendum and is operated by a provincial marketing board. A total of 3 million hogs have been sold on the exchange during a good year. The Alberta and Manitoba provinces in Canada have similar systems for selling hogs.

In the Canadian teletype auction, one control center receives and transmits all sale bids. Each buyer usually has his own teletype connection located at his plant. A Dutch or regressive auction is used,[36] a feature that is believed to accelerate the auction process and bring higher prices.

The Canadian case is a good example of the use of federal legislation essentially to guarantee the existence of a competitive open market. The Virginia teleauction exists without any such enabling legislation. Success depends on its ability to provide a better exchange mechanism than existing alternatives. The Virginia teleauction cannot guarantee the existence of an open market, but it can attempt to create a market more convenient for the seller and buyer to use.

Various forms of legislation have been proposed in the United States as approaches to guaranteeing open markets, e.g., government agency to develop and operate the auction at least until proven feasible, authorization of a marketing board (the Canadian case), a requirement that the auction system be used for a percentage of all trades, and a requirement that a given percentage of all trades go through a conventional central market.

Both the Virginia and the Ontario systems require the existence of a standardized grading system. The product is graded at the farm or at local assembly points by third-party graders (often government personnel). With visual inspection no longer necessary, buyers can make their purchase decisions hundreds of miles from the location of the animal. They make their purchase decisions with the assurance that the grades are an accurate reflection of the animal's quality. Establishment of a system for grades and standards must often be a first step in creating an electronic exchange system.

Another approach to the strategy of maintaining an open market pricing system is to increase the flow of market information to the market participants. (This concept is discussed in more detail in the section on information systems.) Unlike the case with electronic exchange mechanisms, the existing method of pricing would be retained.

Perfect knowledge is one of the requirements for pure competition in the theoretical sense. While the ideal case is never achieved, an increase in relevant market information dispersed in such a way as to help put both buyer and seller on a more equal informational level will increase competition and improve the openness of the market.

An example of the role played by an open-market cooperative in improving the informational base is the use of radio broadcasts by Central Livestock Association, Inc., St. Paul, Minnesota. The cooperative operates a daily radio program aimed at providing cattle and hog producers with the latest trends in market price, volume, and quality movements. Another cooperative information system, Telcot, is discussed in the section on information systems.

Marketing cooperatives use a variety of information systems, both public and private, to help them make more informed marketing decisions. The USDA provides to the public a wide range of market news reports and statistical summaries that are used by all segments of the food marketing system.[37] Interest has been expressed in having the USDA also report closed market transactions such as contract prices for fruits and vegetables.

Marketing cooperatives relying on these widely publicized reports should ask themselves if their information base is adequate. Do they have access to the market information their customers do? Can they afford not to?

The use of market information systems by farmers or their cooperatives does not by itself guarantee an open market system. It should help the agricultural producer come closer to matching the typically larger and more sophisticated information base of oligopolistic buyers. A more open and competitive market should result.

ASSESSMENT. The open market strategy was not a natural concept in the minds of the managers surveyed. They did not initiate discussion of it as they defined their overall goals and objectives. This might be expected since the interviews dealt in most cases with cooperatives marketing products to which they had already acquired a title and on which they had performed some processing functions. A change to an auction form of pricing, for instance, would make little sense to a well-run marketing cooperative. Such an organization through its market power[38] and integrated structure should be able to do better for the farmer than the open market would.

Comments such as "We hadn't really thought about it" (an open market strategy) and "It sounds interesting, it should be looked into" were typical when managers were asked about the desirability of developing electronic exchange mechanisms for some commodities.

Cooperatives marketing a commodity such as grain are currently operating in an open market. But managers of these cooperatives appeared unconcerned about any need to design a system that would guarantee the continuation of that market. Yet their entire marketing approach is presently based on the continued existence of an open market.

Core sample managers did state that one of the reasons for having their cooperative was to maintain a competitive market system for agriculture. In practice, this might be better interpreted as a countervailing power concept, at least for the integrated cooperative. It is unlikely that management of these cooperatives really consider the competitive market in the theoretical sense as a current, important goal. A more important goal for these integrated firms may be to acquire market power equivalent to that of the large agribusiness firms with which they compete or to which they sell.

But probably most agricultural producers do not now have the opportunity to market their product through an integrated marketing cooperative. Most marketing cooperatives such as grain marketing cooperatives are not integrated at all or are integrated only to a very limited extent. In other commodities such as livestock, cooperatives handle only a very minor proportion of the product. In industries such as these, producers and their cooperatives should be concerned about the future viability of open competitive markets. Their ability to receive a "fair price" from the marketplace depends on it.

To some extent these concerns were evident among core sample respondents. Managers were asked to speculate as to the role of cooperatives in situations where vertical integration by cooperatives does not appear feasible. They believed the role of cooperatives would be important in (1) getting member commitments through marketing agreements, (2) pushing for government legislation to maintain an open market, and (3) setting up electronic exchange mechanisms. They were evenly divided on the importance of bargaining. Almost all believed that pushing for government legislation to regulate corporate integrators would not be an important role for cooperatives in the future.

Academic personnel were asked to make the same speculation. They believed the most important role for cooperatives was to encourage member commitment through marketing agreements. Second, third, and fourth in importance were bargaining, pushing for government legislation to maintain an open market, and the use of electronic exchange mechanisms. Pushing for government legislation to regulate corporate integrators was a distant fifth.

Both groups believed the most important role of cooperatives, where integration was not feasible, was to encourage member commitment; the least important role was to encourage regulation against corporate integrators.

The Electronic Auction Exchange as an Approach to an Open Market System. The electronic auction exchange is stressed here because it is probably the most positive and direct approach to maintaining an open market. It is also a rather new concept that can take advantage of the latest technology. To some extent the concepts involved in it are similar to those in other approaches to the open market systems.

The *advantages* of such a system have been pointed out by several authors. In the case of hog marketing, David L. Holder has cited the following advantages: the system would (1) insure a price that accurately reflects supply and demand, (2) increase the physical efficiency in the movement of hogs, and (3) improve pricing accuracy by rewarding each producer for exactly what he produced, by providing him with more accurate market information, and by exposing his offer to a larger number of packers.[39]

However, the electronic exchange mechanism does have potential *disadvantages.* Simply establishing a different method of price making does not by itself assure the existence of a more competitive or more efficient market.

Volume as a percentage of total product marketings is a crucial factor in determining the success of this type of marketing system. Successful operation may require that producers support the system by committing their production to be sold on the exchange for a period of several months. Without seller support, buyer support certainly cannot be expected. If market volume is low, buyers may find it to their advantage to negotiate for price directly with the seller (often resulting in a lower price). Without adequate volume, pricing efficiency will also suffer.[40] It may not even be economically feasible to operate the system. One solution to the volume problem is to push for legislation requiring that a set percentage of the buy orders go through the exchange. Such legislation would, of course, limit the ability of farmers eventually to implement an alternative marketing system.

Another potential problem is the availability of market information. Information generated (although available equally to buyer and seller) may be more advantageous to buyers. Their better knowledge of market conditions may actually enable them to use this new information to the seller's disadvantage.

An auction type market can also end up as a dumping ground for inferior product grades, especially if volume is low. If producers feel their quality products are being inadequately compensated for on the exchange, they will go around the market in hopes of finding buyers who are willing to pay for a quality product. The auction ends up receiving the residual. This has often been the case in auctions such as country cattle auctions and citrus fruit auctions in California.

But the electronic exchange mechanism may well be the strategy that meets with the least resistance from producers, middlemen, and consumers.

It requires little or no financial commitment from producers, although it probably requires some production commitment. Farmers are relatively free to sell when they want and to take their own risks with regard to market price, and they may have greater control over selling decisions if their products do not have to be moved to the central market for selling. To many farmers these qualities combined with the prospect of a more competitive and open marketing system are highly desirable.

Middlemen should also be pleased. They are not faced with any change in market structure on the seller's side. The strategy does not involve the formation of a grower organization capable of negotiating terms of trade from a position of market power.[41] Buyers are advantaged if the auction carries with it a greater pricing efficiency and a greater physical efficiency in the movement of product from farm to plant. They may be able to select among a much wider range of product than was possible before. One disadvantage, however, is that buyers may be forced into greater competition with other buyers.

Society as a whole should also benefit. The improved physical coordination would result in savings. The electronic exchange mechanism should also allow a more even distribution of economic power. Consumers could expect cheaper food from a more competitive market system while at the same time maintaining the independence of the individual family farmer.

The attractiveness of this strategy from the standpoint of society is especially significant in today's political environment. Inflation has popularized charges that regulation increases consumer prices. A return to the open market without government regulation would be well received by many.

The attractiveness of the electronic exchange mechanism for cooperatives may be looked at in two ways. *First,* it may be a special opportunity for cooperatives. Farmer support is essential for this strategy. The close tie between cooperatives and their patrons gives cooperatives a natural advantage. In addition, the rapid development of technology may present significant opportunities for those willing to be innovative in taking the initiative in this area. *Second,* cooperatives may have a responsibility to act to perpetuate the open market in some commodities. The potential for government intervention in the pricing of agricultural products will become greater as markets move further away from competitive conditions. Cooperatives can act now to improve market access for producers. If they do not, government policy, motivated by the concerns of society, may do it for them. If this happens, farmers may have little say in determining how they will be able to market their products.

Many economists favor a competitive market. They believe that the purely competitive situation is the preferred market for farmers. The electronic exchange mechanism is one method of achieving this. But not all economists or farmers are so sure that farmers would fare best in such a market. Farmers may well be substantially better off if they can alter their position as price takers in the market. Insofar as they can build some degree of market power through their cooperative organization, they may be wise to set the integration strategy as their ultimate goal.

Many of today's cooperatives, whether they realize it or not, are operating with an open market strategy. Such a strategy requires the existence of an open market as discussed earlier. These cooperatives and their members must recognize that continuation of the open market strategy will require an awareness of the forces working to close off their markets. Pushing for such innovative approaches as the electronic exchange, improved government information systems, or more regulated markets may well be crucial to their cooperative's future existence.

5 Organizational Strategies

THE THREE general strategies identified in Chapter 4 are conceptual ideas. Operationally, they can be put together in a variety of ways. In fact, their success may depend as much on the wisdom of the management and farmers involved and the way the particular strategy is designed as on the choice among the three concepts. General strategies can use one or more of several organizational strategies—horizontal integration, diversification (and conglomeration), multi-cooperative organizations, and joint ventures.[1]

Horizontal Coordination

Horizontal coordination has been widely used as an avenue of growth by cooperatives. In addition to horizontal expansion internally, cooperatives can grow externally through merger, federation, or other multi-cooperative organizations.[2]

This section draws mainly on research and observations made in the field by the research team. Horizontal integration is a broad subject well covered in literature, and the research team did not seek input on the subject directly from managers of cooperatives studied.

CONCEPT AND THEORY. Horizontal coordination through external means involves the uniting under common ownership of two or more businesses engaged in the same level of production or sale of product. It enlarges the basic function of the firm without changing that function.

Horizontal coordination also consists of expansion at the same production or marketing level through internal means. The difference with horizontal coordination in this internal case is that growth occurs solely within a single firm. New facilities are constructed or new services are added to serve the industry at the same marketing level.

Growth in the horizontal dimension is closely associated with, and is often essential for, vertical or diversified growth. The history of cooperative growth has clearly indicated this close relationship. Joseph G. Knapp writes that, as horizontal growth occurs,

> volume is built up, sufficient for large-scale operations, making it possible to strengthen the power of the organization through broadening its services by diversification, or by expanding vertically to take on such functions as wholesaling, transportation, research, advertising, processing, or manufacturing. As the efficiency of the organization is increased by vertical integration the advantages of membership are increased and more members are added, thus resulting in more horizontal integration.[3]

A certain amount of horizontal growth is often an initial step in the organization of any cooperative—usually a prerequisite before the more advanced vertical integration strategy can be achieved satisfactorily.

A primary means of horizontal coordination for local cooperatives has been the adoption of a federated form. The federated cooperative has been dominant among supply cooperatives, but it is also used by marketing cooperatives, especially those marketing grain. Local cooperatives organized as a federation permit the grouping together of many units performing basically the same functions. Ownership remains at the local level, and local control is retained over most operations. The horizontally organized group of locals thus forms a regional cooperative. The regional can perform functions such as manufacture of farm supplies, centralized marketing or exporting, advertising, or processing—functions that could not be effectively performed at the local level. The local-regional cooperative combination might now be called vertically integrated.

Cooperatives also use the centralized form of organization allowing direct farmer membership, with local units being managed by a regional cooperative. Centralized organizations can have basically the same properties of horizontal and vertical integration as the federated organization except for the lack of local control. Potential coordination among horizontal and vertical elements is much greater for the centralized than for the federated organization because of this control.

The interregional cooperative is another example of both horizontal and vertical integration. Membership consists of a horizontal federation of regionals which become vertically integrated forward or backward to perform a specific function. CF Industries, for instance, is made up of nineteen regional cooperatives having for the most part agreed to give up their own independent manufacturing of fertilizer and turn the entire operation over to CF Industries. The horizontal grouping of nineteen separate businesses forms an assured market and capital base for the managerial talent and facilities investment needed to integrate backward into plant food production.

MOTIVATION. What are the motivating factors for horizontal coordination? Firms, cooperative or noncooperative, may engage in horizontal coordination solely to satisfy the urge to grow. Usually other benefits exist, and they include the following:

1. Opportunity for economies of scale in assembly, processing, or marketing.
2. Opportunity to utilize more fully existing managerial expertise or to acquire new managerial expertise.
3. Opportunity for plant specialization.
4. Opportunity to eliminate duplication of services, facilities, and other resources.
5. Opportunity to increase market power to countervail existing market forces.

These reasons are quite important in considering the general strategy of integration and coordination. Two studies of cooperative combinations through merger have been conducted by Bruce Swanson and by Leon Garoyan and Gail Cramer.[4] Both provide interesting insights into the achievement of merger objectives.

Swanson showed that three years after the merger the reorganization had resulted in a "drag effect on the margins earned by the acquiring component, but [had been] highly stimulative in a positive fashion in terms of the margins earned by the acquired organization."[5] Start-up expenses were evidently responsible for a significant part of this drag on net margins. Swanson also did a detailed financial analysis of postmerger firms. He found that the liquidity position of the reorganized cooperative was considerably stronger than it had been in either firm before the merger.[6]

Garoyan and Cramer found that although all mergers studied had been motivated to attain economies of size, only 40 percent of the merged facilities had achieved it five to twelve years later, apparently due to their inability to dispose of excess acquired facilities. The researchers believed the problem could have been remedied in many cases through more adequate premerger planning.[7] In contrast, almost all the smaller acquired cooperatives had achieved economies of size over the same periods. They also found that, during the same periods, acquiring cooperatives attempting to increase their marketing or bargaining power through merger "failed to attain their objective because of changes in technology, supply response of producers, and the structure of the market."[8]

HISTORICAL TRENDS. While cooperative horizontal coordination can take a variety of forms, historical data exist only for the merger form of coordination. Mergers do, however, serve as a good measure of horizontal coordination among cooperatives. Swanson found that out of a total of 787 cooperative reorganizations occurring during the important merger period, 1960–69, 666 or 85 percent involved horizontal growth.[9] By far the largest number of these reorganizations during the 1960s occurred among dairy co-operatives, representing 365 of the 787 total. Grain and livestock coopera-tives were ranked second and third with 96 and 36 reorganizations respec-tively.[10] Horizontal reorganization usually occurred at the raw product pro-curement and first-stage processing levels. No statistics are available for the 1970s, but indications are that the merger movement has abated from the rapid pace of the sixties.

Cooperative merger statistics prior to 1950 are incomplete. Available data indicate 172 mergers among cooperatives during the 1920s, 217 mergers during the 1930s; and 359 during the 1940s.[11] In the 1950–59 period, merger activity declined to 222 reorganizations.[12]

Garoyan and Cramer in studying the relationship between cooperative mergers and general business conditions over the 1920–64 period found that "cooperative mergers have occurred during periods when stock prices (ex-pectations) are high and when farm income is favorable—cooperative merger activity is not closely associated with the business cycle, nor do co-operative mergers occur when other business firms are failing."[13]

RELATED ISSUES. Most cooperative managers in the core sample said that they saw more legal constraints for cooperatives through horizontal expan-sion than through vertical expansion.

Horizontal merger combinations among business firms in the past have received the greatest amount of antitrust attention, especially when the combination is thought to result in a reduction of competition.[14] Market share measures before and after the combination are important considera-tions during legal proceedings. However, no across-the-board market share guidelines are used by regulatory agencies in determining the legality of a merger, although there are some guidelines used to signify which mergers are likely to be challenged. Antitrust laws seldom block mergers undertaken to save a small firm in danger of going out of business.[15]

Cooperatives, especially with large market shares, are under closer scrutiny with regard to their merger plans than they have been in the past.[16] Marketing cooperatives, because of the Capper-Volstead Act, *may* have more flexibility with regard to horizontal mergers than do supply coopera-tives. The act would appear to allow two marketing cooperatives to achieve a merger despite the Clayton Act simply by allowing the formation of a new cooperative consisting of the members of the original two. But this path to merger is by no means clear nor is it free from the possibility of antitrust problems. A merger thought to restrain trade or diminish market competi-tion would be subject to legal challenge. The Capper-Volstead Act explicitly disallows cooperative associations to monopolize or restrain trade to such an extent that prices are unduly enhanced. Furthermore, the Justice Depart-ment or Federal Trade Commission might challenge any horizontal merger occurring between two large regional cooperatives.

Another issue affecting horizontal coordination relates to competition among cooperatives. Competition among cooperatives serving the same geographical area can sometimes be justified. A new cooperative moving into an area can bring new vigor, new thinking, and new services. This may be especially true if no competition existed previously. Unfortunately, competition among cooperatives can also result in unnecessary duplication of facilities and overhead.

ASSESSMENT. Swanson concludes that the future potential for cooperative mergers is considerable. This prediction is based on the large number of cooperatives that had not been involved in merger activity as of 1969. Although merger activity was substantial during the 1960–69 period, almost 91 percent of the cooperatives existing were not involved in any reorganization activity that included another firm.[17] The lack of past merger activity is not in itself an indication of future activity. But the future competitive environment for cooperatives is likely to force more mergers. Conglomerate agribusiness firms; environmental, health, and safety regulations; and higher cost facilities all will make it necessary for cooperatives to group together to achieve a balance of power and to achieve economies of size. Legal constraints will probably cause more mergers among small cooperatives than among large regionals.

Results of the Garoyan and Cramer study indicate that successful cooperative mergers mandate careful planning of the postmerger organization, especially in the area of facilities use. Cooperatives attempting reorganization for the purpose of enhancing market power should carefully study the structure of their market as well as the likely producer supply response.

Assuming that future horizontal mergers among large cooperatives will be given increased public scrutiny, it may also be wise to consider alternative forms of external horizontal coordination.[18] Cooperative federations represent one widely used alternative. The regional cooperative can supply the services possible only through a grouping together of smaller cooperatives, yet no merger is required and locals retain their autonomy and control over local level functions best handled at the local level.

Currently, considerable interest exists in using the interregional cooperative to achieve horizontal and vertical coordination among regionals. One manager said that the trend over the next five years will not be to merge, but to keep separate entities and to form interregional operations. An interregional such as National Cooperative Refinery Association, McPherson, Kansas, is able to build the petroleum refining capacity needed to serve thousands of farmers affiliated with the six regional members, yet no merger is required. However, conflicts in management philosophy sometimes occur between interregional and regional organizations.

Two other forms of horizontal coordination are available to cooperatives—the "marketing agency in common" specified in Section I of the Capper-Volstead Act and the exchange of market information specified in the Agricultural Marketing Act of 1926. However, the legal interpretation of both of these is not clear and is subject to considerable debate.

Cooperatives wishing to use either of these provisions are strongly advised to seek competent legal help.

Diversification (and Conglomeration)

Diversification is a widely used business practice. This organizational strategy can be used in conjuction with any of the three general strategies to aid cooperative growth. It is most often thought of as a part of the integration and coordination general strategy.

CONCEPT AND THEORY. Diversification can be defined as a departure from current product or service lines and present market structure. According to H. Igor Ansoff, diversification requires "new skills, new techniques, and new facilities." Growth in this direction usually leads to physical and organizational changes representing a distinct break with past business experience.[19] Philip Kotler defines diversification by breaking it down into three classifications: concentric, horizontal, and conglomerate diversification.[20]

Concentric diversification occurs when a company adds new products or services that have technological and/or marketing synergies with the existing product line. Normally these products would appeal to new classes of customers. Sunkist has practiced concentric diversification by developing expertise in the product research area for the development of its own citrus products and is now in a position to directly market the results of its product research.

Horizontal diversification is the addition of new products or services appealing to present customers though technologically unrelated to the present product line. A good example would be a cooperative such as Mid-Continent Farmers Association (MFA), Columbia, Missouri. This organization not only produces feed for sale to farmers through local cooperative retail stores but also sponsors a separate mutual company to sell life insurance to these farmers.

Conglomeration as a form of diversification carries fewer restrictions than do the other two forms. To achieve conglomerate diversification, a company enters new product areas that are related to existing areas only with respect to managerial and financial functions. There are no product development, purchasing, or marketing similarities. Products or services are usually sold to entirely new classes of customers. Typically, a company achieves the conglomerate status through merger or through acquisition of another firm whose product line is completely dissimilar.[21]

CURRENT SITUATION. Growth through diversification has been extensive in the American economy. In 1949 more than two-thirds of the top 500 United States industrial corporations were committed chiefly to one primary business area. Twenty years later, two-thirds were following strategies of broad diversification.[22] However, Albert G. Madsen and

TABLE 5.1. Sideline activities of marketing, farm supply, and related service cooperatives for 1950–51, 1960–61, and 1973–74

Type of cooperative and its sideline activities	Period		
	1950–51	1960–61	1973–74
	Percent engaging in indicated sideline activities		
Marketing			
Farm supply	60	65	58
Related service	49	64	64
Farm supply			
Marketing	22	26	26
Related service	21	53	49
Related service			
Marketing	20	5	10
Farm Supply	40	44	28

Source: Martin A. Abrahamsen, *Cooperative Growth Trends, Comparisons, Strategy,* FCS Information 87, FCS, USDA, Washington, D.C., Mar. 1973, p. 24; and Swanson and Click, *Statistics of Farmer Cooperatives,* p. 11.

Richard G. Walsh found in 1969 that only 8 of the largest 200 corporations qualified as true conglomerates.[23]

The most obvious example of cooperative diversification is the combining of marketing and farm supply businesses. In 1973–74, 58 percent of marketing cooperatives were handling some farm supplies. Sixty-three percent were handling services related to marketing. Among farm supply cooperatives the percentages were much smaller—only 26 percent handled marketing functions and 49 percent handled related services (Table 5.1).

All the core sample cooperatives with the exception of Riceland foods were providing both marketing and farm supply functions.[24] Most managers surveyed believed that cooperatives would become more diversified in the next decade.

Diversifying cooperatives have put together a multitude of agriculturally related activities under one umbrella. MFC Services is quite small compared to diversified organizations like Farmland Industries, Inc., Land O'Lakes, Inc., Gold Kist, Inc., yet it provides an example of the level of diversification common in small regional supply-marketing cooperatives (Table 5.2).[25] MFC is also developing new local cooperatives and services to satisfy the needs of Louisiana farmers. Among other things, MFC maintains a rice mill. The cooperative is also expanding into foreign marketing through its international marketing division, Agri-Business International.

Diversification may lead to development of more full-service centers at local levels. A full-service cooperative would supply all producer input needs, a market for all their products, and would even provide a coordinated source of capital. Other services such as insurance and travel planning are possible. The cooperative having reached this level of diversification also achieves a more balanced risk exposure.

Ronald D. Knutson has stated that the full-service concept would satisfy the producers' increasingly complex needs in the areas of speciality services, input availability, market access, risk reduction, and capital.[26] Full-service concepts in the highly diversified marketing-supply coopera-

TABLE 5.2. Marketing and supply operations of MFC Services federated cooperative

Company division	Function
Egg operations	With producers: breeder flocks hatchery contract egg production feed mills poultry, equipment, and supplies egg-grading facilities processing plants marketing and distribution
Livestock operations	With producers: livestock feed manufacturing poultry feed manufacturing animal health services livestock equipment
Poultry operations	With producers: hatchery contract poultry raising poultry processing poultry marketing cotton warehousing grain marketing
Crops and agronomy operations	To member cooperatives: distributes plant food distributes seeds distributes agricultural chemicals distributes general farm supply items and equipment

tives are already evolving, especially among such cooperatives as those in the core sample.

MOTIVATION. Kotler writes that diversification growth is a logical path for a company "if the core marketing system does not show much additional opportunity for growth and profit, or if the opportunities outside the present core marketing area are superior."[27] Other writers go further; Ansoff stresses the following reasons for diversification:

1. As compensation for technological obsolescence in original product lines,
2. As a means to distribute risk,
3. As a means to utilize excess productive capacity,
4. As an avenue for reinvestment of earnings, and
5. As a means (in the case of merger) to obtain top management.[28]

Diversification is also entered into because of the legal restrictions often placed on horizontal growth. Diversification is typically not aimed at market control or concentration—two goals often creating legal problems for horizontal growth. Diversification can also bring with it the advantage of an increase in economic power even though specific market control is not attained.

A typical rationale for diversification in an agriculturally related business is to acquire product lines that will even out seasonal income flows and distribute the risk of adverse fluctuations in seasonal commodity prices.

Cooperatives also have considerable opportunity for diversification to achieve a balanced risk exposure. Several different products may be marketed or several kinds of supplies may be handled. Poor marketing receipts can be balanced by the steadier income derived from selling farm supplies. One manager of a diversified supply-marketing cooperative stated that "with the high risk and low margins in marketing [farm products] you have to have staying power." He was speaking of a meat packing venture backed up by a profitable farm supply business.[29]

An important difference between reasons for diversification in noncooperative firms and those in farm cooperatives lies in the owner-patron aspect of cooperatives. A cooperative may also be able to take advantage of the usual reasons for diversification. But the most important reason will often be the additional service the diversification can give the farmer members. Members desiring a new service or product may believe that insufficient competition exists among firms currently serving them. The cooperative may need to diversify to correct the situation. Cooperative diversification may also originate to take over the functions of a company no longer serving a particular area. Agway, basically a supply cooperative, set up the Curtice-Burns/Pro-Fac joint venture for this reason. The need was to provide vegetable processing facilities to growers who were in danger of being left without processing allocations. Agway helped fill that need and helped set up a farm-to-consumer marketing channel and, through its partial ownership of Curtice-Burns, added a profitable diversification to its farm supply business.

Burson, executive vice-president of Gold Kist, described diversification at Gold Kist as follows:

> We are adaptable to change. We started out with cotton but got involved in other things. Members requested market help and then supplies such as fertilizers, chemicals, and insecticides. Then we merged with a pecan marketing group called Gold Kist and gradually started using the name. When we realized that about 1 percent was cotton, we changed the company name to Gold Kist.[30]

Members are likely to consider diversification in their cooperative as a means of adding products or services that will enhance their own farm business. They are not likely to be particularly concerned about the rate of return on the investment, unless the diversification damages the financial strength of the cooperative. In contrast, stockholders of noncooperative firms see diversification only in terms of what it will do to the value of their stock or dividends.

Cooperative management is caught in a dilemma. Their concern must be not only for the growers (and for equity among them), but it must also be for the long-range viability of the cooperative. Even though the cooperative can often justify diversification (even a diversification that would not provide a sufficient rate of return for a noncooperative firm), it must not endanger its ability to serve producers in existing ways as well.

POTENTIAL PROBLEMS. A common mistake made by the diversifying firm is to acquire a company so totally unrelated to its present competence or needs that management cannot effectively handle the new business. A Harvard study of 250 large diversified firms found that the highest level of economic performance occurred among firms using "controlled diversity—diversification that builds on the central skills or strengths of the company."[31] The study suggests that diversification built solely on the financial resources of the company as a base can lead to severe managerial problems. A *Wall Street Journal* article pointed to an awareness of this problem when companies consider acquisition or merger, saying "companies appear far more selective than in the 1960's. Wary of spreading themselves too thin, they insist on buying only companies that fit their corporate strategy."[32]

Cooperatives taking the diversification route need to recognize the risk not only of neglecting their company's major business strengths but also of neglecting the agricultural strengths of their members.

Growth through diversification can generate some other problems for cooperatives that are not faced by noncooperative firms. One such problem is conflict among members. For example, farmers selling soybeans may question the need for new cooperative investment in programs designed to assist hog producers. The new program will be in serious jeopardy if it threatens to become a drain on capital committed to soybean processing and marketing. Member conflicts also slow the decision-making process in a diversified cooperative. The constant need to seek compromise may leave the diversified cooperative unable to adjust rapidly to its changing economic environment.

Cooperatives often face nonmember problems as they diversify. To obtain sufficient volume to be profitable, a new business venture may require that the cooperative begin conducting substantial amounts of business with nonmembers. Because of tax legislation and the Capper-Volstead Act, cooperatives are constrained in their diversification to ventures that can be largely supported by member business. Cooperatives must be careful to keep the interests of members dominant.

The problem of member involvement and control may be accentuated in a large diversified cooperative. Individual producers may not retain the same loyalty to a diversified cooperative as they would to a single product organization. Nor can they be expected to be as knowledgeable or to participate as effectively in the diversified business. One solution is to tie members directly into commodity programs as Land O'Lakes has done with its turkey producers. The diversified cooperative can also make better use of its board member talent by forming special study committees capitalizing on the strengths of individual board members.

CONGLOMERATE GROWTH. Cooperatives, because of certain limitations on their activities, may be restricted from conglomerate diversification. Marketing cooperatives, under the Capper-Volstead Act, are restricted to the marketing of products of "Persons engaged in the production of agricultural products." Both marketing and supply cooperatives, if they wish to be eligible for Section 521 status, are restricted in that member sales must be greater than nonmember sales.[33] Usually the cooperative's articles of incorporation include clauses restricting cooperative activities to agriculturally related areas directly serving agricultural producers and farmer members. But most importantly, cooperative purpose and objective statements approved by cooperative boards contain similar statements. Examples include statements such as: "The purpose is to improve the economic well-being of members by supplying products and services for farm production and marketing farm products," or "The purpose is to engage in any activities in connection with the marketing or selling of the agricultural products of its members and other patrons."

Although cooperatives, as mentioned, are restricted both legally and philosophically in their ability to extend their operations beyond agriculture, some cases of cooperative diversification do come close to the conglomerate form while still remaining in agriculturally related fields. Farmland Industries, for example, has diversified from the simple marketing of farm supplies into petroleum and oil exploration; petroleum refining; pipeline operation; fertilizer manufacture; financial services including insurance, lending, and auditing; meat packing; and marketing of processed products.

Conglomerate diversification, according to the traditional definition, would require that a cooperative market products or provide services having no relationship to agriculturally based production. Marketing strategy for such products would probably be aimed at nonfarm customers. The cooperative would serve only as a base of managerial talent or of financial resources. Such activities would put cooperatives in serious danger of losing their member support. The researchers found no organizations of this type.

Diversification of a very broad form may under certain circumstances be possible and perfectly legal for cooperatives. A cooperative can, for example, form a wholly owned noncooperative subsidiary. It can also legally handle some nonfarm business. Both of these practices are common among cooperatives.

Cooperative managers in this study were asked whether cooperatives would have to find ways to branch out into nonagriculturally related areas to enable them to compete with conglomerate organizations. Almost every respondent believed that cooperatives would not need to get into nonagricultural fields. The majority of the university experts polled responded in the same manner.

Cooperative managers saw plenty of opportunities for diversification within agriculturally related fields. One manager said, "Diversification gives us strength, but we can't see diversification going beyond agriculture." Another manager saw diversification beyond agriculture as a dangerous drain on financial resources needed to serve agricultural production. Managers of these cooperatives seemed well aware of the fact that their primary mission was to strengthen the position of the farmer. They could not see that diversification into nonagricultural areas was necessary to satisfy that mission.

Cooperatives, nevertheless, may be at a disadvantage in their ability to compete because of the conglomerate activity of firms operating simultaneously in both agriculturally and nonagriculturally related industries. This conglomerate activity often extends even further with the multi-national companies dealing in the products of many countries. Willard F. Mueller writes:

> If a large multi-market firm occupies substantial market positions in some of its product or geographic markets, it may use the profits derived there to subsidize its attainment of market power elsewhere.[34]

Conglomerates with this kind of power may be a major obstacle to the ability of cooperatives to perform their economic role.

ASSESSMENT. Diversification is important to the agricultural cooperative just as it has been to the individual farmer. Cooperatives, in fact, have a responsibility to diversify if they are to meet the needs of their diversified producer-members. Respondents believed that diversification is not severely limited because of the inability or lack of desire of the cooperative to diversify into areas outside agriculture. The range of activities left open to cooperatives within the agricultural area remains quite broad. The important point for cooperatives is that growth through diversification be recognized as an organizational tool for better serving the needs of farmer members. It should be recognized, however, that in some cases the single commodity marketing cooperative may be better suited for enhancing the economic welfare of the producer.

Some cooperatives do have a diversified structure that borders conglomeration. Whether or not they actually are conglomerates is a question of definition. In any case such cooperatives should be cautious in their approach toward broad diversification. There may be serious question from the standpoint of society and the law whether a cooperative operating with a certain amount of federal sanction should use this base as a launching pad for diversified activities that increase its economic power and reduce the competitiveness of the marketplace. The issue becomes important as cooperatives move into activities that are only remotely tied to production agriculture.

John Kenneth Galbraith pointed out in *The New Industrial State* that the motivation toward growth and diversification, whether by merger or internal development, is something inherent in the modern corporation.[35] There is no reason to believe that cooperatives are not subject to this same motivation. But the cooperative, because of its owner-patron tie and its federal sanction, is not likely to utilize the diversification concept to the same extent as the noncooperative firm.

Multi-Cooperative Organizations

The organizational strategy of multi-cooperative organizations is an important and highly diversified group of arrangements, recently attracting attention.

CONCEPT AND THEORY. A multi-cooperative organization or cooperative venture may be described as an association between, or federation of, two or more cooperatives organized for the purpose of conducting specific functions, but with the identities of the participants remaining apart from their participation in the venture. It is a joint venture but excludes noncooperative organizations and normally includes many more firms than in a joint venture.

Such organizations divide generally into two groups—commercial and noncommercial. The commercial type organizations would include those who actively participate in one or more of the following: manufacturing, procurement, processing, wholesaling, or marketing of farm or farm-related supplies and products. The regionals, interregionals, and any federation of interregionals would fall into this grouping.

Noncommercial organizations would include educational, trade, research, and other related organizations. The purpose of these organizations is to carry out specific functions that are necessary for maintaining or improving the cooperative way of doing business.

These organizations may be a valuable aid or tool in any of the three general strategies. Historically, the noncommercial organizations have dominated, but more recently the commercial organizations, especially the interregional, have been developed and promise to become important.

Business management has long recognized that some needs of businesses are best met by joint efforts. Trade associations and many other efforts exist. Historically, however, public scrutiny has sharply restricted what noncooperative companies can do by concerted effort. Today, cooperatives have more latitude to engage in such joint organizations than do noncooperatives.

Multi-cooperative organizations in the United States have been used almost as long as formal agricultural cooperatives. In the late 1860s, the Grange movement promoted the idea of agencies.

> An agent or agency of a local or of several Granges acted for all members. Orders for supplies were lumped together, and volume deals were made with jobbers and manufacturers at savings over the usual retail prices. Similarly, savings were effected by the pooling of farm products for shipment to the large markets in order to bypass the usual toll taken by local commission men and dealers.[36]

The major advantages of multi-cooperative organizations are:

1. They help to satisfy an economic need. Multi-cooperative organizations were born out of an economic need of farmers to develop countervailing and bargaining power through pooling of capital resources and/or commitments to certain agreements. In general, certain areas exist (such as obtaining raw materials) where individual cooperatives by themselves do not have the resources to compete. They are not large enough to take advantage of the needed economies of size or to accept the risk involved in carrying out a particular activity. Multi-cooperative agreements provide individual organizations the advantages of economies due to large size.

2. They often reduce possible risks and economic losses for individual organizations. By joining together, individual cooperatives need to put up less capital and share the risks of financial losses.

3. They allow participating cooperatives fuller use of their present resources. In most cases, duplication of resources and services can be eliminated. Thus a cooperative does not need to spread itself too thin by devoting part of its managerial expertise to areas encompassed by management hired especially to carry out the particular venture. A multi-cooperative organization usually provides the resources necessary to hire specialized management expertise needed for the special purposes of the agreement.

Potential problems with multi-cooperative organizations or their lack of success can be contributed to several factors:

1. Some managers were concerned about loss of identity or political power if their organization joined with other cooperatives. Managers mentioned philosophical conflicts that result in inaction to agree on what should be done. Equity problems can arise with big cooperatives working alongside the little cooperatives. One manager also expressed the fact that there was not much glamor for his cooperative in going together. "A certain amount of competition among the regionals is important and you need to retain substantial individual identity if you are going to have producer support."

2. Individual organizations may be unwilling to give up certain geographic advantages. As one manager expressed it, "One of the hangups in any commonality would be to what degree do we have the right to bargain away inherent geographic advantages that our producers might have in order to put something together that would be good for farmers in general." "There is," voiced another manager, "a big difference between what actually is and the theoretical ideal of being one big cooperative family."

3. Some managers were concerned about the commercial type organizations moving either too fast or too slow. One manager was distressed with the potential hazards of interregionals. In some ways, he believed that people are becoming too enthusiastic about the idea of interregionals. "The hazards," he explained, "are that some of the interregionals may start running faster than the [financial] resources justify. Another hazard is that regionals or interregionals in the process of pulling in people may get people with noncooperative backgrounds, with great talent in one area but the people do not have or do not know the purpose of the organization and do not know the politics of maintaining a member support base." He said that he was not sure about the

> right balance between managing the power that's available in the outside world (that is, hiring managers with expertise in business management) and at the same time maintaining a feeling of philosophical commitment to the purposes of cooperative organizations. You may find that you are a big enterprise that is no different than any other business out there.

A few other managers expressed the view that multi-cooperative organizations in certain ventures (foreign trade, for example) are too slow to make decisions. As one manager explained it, "You can't get a mutual meeting of the minds when you get a number of cooperatives going together."

4. Participants may lack mutual trust and the necessary commitment needed for the organization. Commitment is one of the major factors to cooperation. Ernest V. Stevenson, Executive Vice-President and General Manager of Illinois Grain Corp. and FS Services, Inc., in discussing grain exporting through a national export cooperative, stated that

> a national cooperative should have the assurance that it can cover those [foreign grain] sales with commitments from regionals, who in turn need commitments from locals, who in turn need commitments from farmers. It comes full circle. For farmers to control their own destiny in marketing—and for farmers to get the full benefits from cooperation—there needs to be willingness to make commitments throughout the system.[37]

The general belief among managers interviewed was that there was "greater willingness in putting efforts together" and that the mood for multi-cooperative organizations is greater now than it has ever been. "If something cannot be done economically by one organization, it ought to be done by a group." "There should be more joint cooperation among cooperatives. There is going to have to be." "We must choose the cooperative level that can best provide the service." "We feel that it is better for co-ops to go together since they are spending the farmers' money no matter how they go. There is already enough competition with private industry so that you don't have to worry about decreases in co-op competition." Mention was made of the need for farmers to become more basic in the production of petroleum and fertilizer, and that these ventures require a cooperative organization at a level where capital can be made available without too great a risk for individual cooperative organizations involved.

METHODS OF APPLICATION. Multi-cooperative organizations are organized at different levels. Many local cooperatives are members of one or more federated regional cooperatives. Regional cooperatives also have a number of areas where interregionals provide economic and other advantages. The regional cooperatives can then better serve the local cooperatives. Cooperation among regionals has taken place in a number of areas: fertilizer production, petroleum discovery and production, farm supplies procurement and manufacturing, transportation and distribution, research, legislation, and education.

Two good examples of interregional cooperatives in the commercial group are CF Industries, Inc., Long Grove, Illinois, and Universal Cooperatives, Inc., Alliance, Ohio.

CF Industries, Inc., is a federation of nineteen regional cooperatives. Two of the members are cooperatives in Canada. Each member of CF has an input into policy decisions by having one representative on the board of directors and one representative as a shareholder. In some cases, the same person serves as director and representative.

The purpose of CF Industries is to supply its members good quality fertilizer and other related services economically. The interregional relies heavily on obtaining noncooperative management expertise to carry out its operations. In fiscal 1977, its sales were $633 million from the sale of 7.6 million tons of plant food—nitrogen, phosphate, and potash. Pretax, prepatronage earnings for 1976 and 1977 were $77.1 and $47.6 million respectively. Patronage refunds to member companies amounted to $70.9 million in 1976 and $7.3 million in 1977. The amount of patronage refunds in 1977 was low owing to a change in board policy on amount to be refunded. The policy was changed for investment tax credit purposes.[38]

Universal Cooperatives, Inc., is "a purchasing, manufacturing, and merchandising interregional cooperative which manufactures, processes, or purchases many types of farm and home supplies. Universal is owned by 38 regional, agricultural, and/or consumer cooperatives located throughout the United States, Canada, and Puerto Rico."

Universal Cooperatives is the result of a merger in 1972 between National Cooperatives of Albert Lea, Minnesota, and United Cooperatives of Alliance, Ohio. Both interregional cooperatives "were founded in the 1930's to utilize pooled purchasing power for the benefit of the nation's hard-pressed farmers."[39]

Universal's sales in fiscal 1976 were nearly $217 million. For a ten-month fiscal year in 1977, sales were nearly $210 million and savings amounted to $5.3 million.[40] During 1977 Universal Cooperatives moved its headquarters from Alliance, Ohio, to Bloomington, Minnesota.

A number of noncommercial cooperative organizations exist among regional cooperatives. In the area of cooperative legislation, education, and research, for example, cooperatives have been able to take advantage of commonalities through the following organizations: The National Council of Farmer Cooperatives (NCFC), American Institute of Cooperation (AIC), and Cooperative League of the USA—all headquartered in Washington, D.C.; Cooperative Research Farms (CRF), a feed research network headquartered at Charlotteville, New York; and Farmers Forage Research (FFR), a plant-breeding cooperative headquartered in Lafayette, Indiana.

The NCFC is a nationwide association sponsored by agricultural cooperatives and cooperative state councils to represent farmers and their cooperatives on national and international matters. Staff members of NCFC work closely with Congress, the executive branch, and the federal regulatory agencies "to create and maintain a favorable political and economic climate in which farmers and their cooperatives can do business." More than 75 percent of United States agricultural cooperatives or some 5,800 individual cooperatives with nearly 3 million farmers are represented by NCFC through regional cooperative membership.[41] Funds for operation of the council are obtained through membership dues. Individual cooperative members' dues are based on their annual business volume.

The AIC is a nationwide association of agricultural cooperatives for cooperation education. "Through cooperation among cooperatives, AIC seeks to stimulate sound, constructive thought on cooperative principles, practices, operations, accomplishments and opportunities. Its educational programs are designed to promote a better understanding of what cooperatives are and what they can and cannot do." Membership in AIC includes local, regional, and interregional cooperatives; cooperative state councils; and other farm organizations. Approximately 1,200 organizations financially support AIC.[42]

Cooperative League of the USA, a national federation of cooperatives, is "devoted to the idea of cooperation by all people who have economic and social needs and the desire to work together in good will."[43] The Cooperative League is involved in cooperative education, information and public relations, development, leadership, and other services.

Two examples of joint research efforts are CRF and FFR. CRF is a federated organization owned by 21 regional cooperatives. CRF includes seven livestock, dairy, and poultry farms in the United States. Research is carried on by over fifty full-time research scientists in the areas of feed and nutrition, management, breeding stock, and marketing.

Research information developed by CRF is available to all members. Each member has an opportunity and obligation to use the research in its own programs. How it is used by each organization is independent of CRF and other members.

FFR is a federated organization owned by fifteen cooperatives. FFR researchers do plant breeding work with a number of forage crops, such as bluegrass, alfalfa, and fescue. Projects in FFR are carried on to develop new varieties or varieties that are resistant to certain insects and diseases.

Other functions for the economic benefit of farmers are presently being carried out by multi-cooperative organizations (Table 5.3).

POTENTIAL PROBLEMS. Certain limitations or constraints to multi-cooperative organizations (especially at the interregional level) exist and must be overcome if such organizations are to be successful.

Managers and directors must recognize when it is economically beneficial for their own organization to join in a multi-cooperative organization. This means giving up the drive for individual cooperative power for the good of members or of farmers in general. This will require compromise and a balance between coordination with multi-cooperative groups and their own cooperative's identity.

Benefits may not be equally divided. Part of this is due to some cooperatives having a lot more to offer the organization than other member cooperatives. Thus their return is not as great as those cooperatives that start with less.

There appears to be momentum to greater use of interregionals in the future. So far, interregional cooperative organizations have received no

TABLE 5.3. Examples of commercial and noncommercial multi-cooperative organizations by major functions performed

Major function	Name of organization
Commercial Type Organizations:	
Cotton marketing—domestic and foreign	AMCOT
Crop services	Servi-Tech, Inc.
Crude oil exploration	International Energy Cooperative, Inc. (IEC)
Crude oil refining	National Cooperative Refinery Association (NCRA)
Distribution	Ag Foods, Inc.
Farm supply purchasing and manufacturing	Universal Cooperatives, Inc.
Farm ingredient purchasing and distributing	CF Feeds, Inc.
Fertilizer purchasing and manufacturing	CF Industries, Inc.
Fruit and vegetable exporting	California Valley Exports
Future trading	Illinois Cooperative Futures Company
Grain exporting	Farmers Export Company
Grain marketing	St. Louis Grain Corporation
Seed corn production and processing	Cooperative Seeds, Inc.
Transportation	Agri-Trans Corporation
Noncommercial Type Organizations:	
Crop and livestock research	Cooperative Research Farms (CRF)
Education, information, public relations, and development—any type of cooperative	Cooperative League of the U.S.A.
Education, information, and public relations	American Institute of Cooperation (AIC)
Forage research	Farmers Forage Research (FFR)
International cooperative assistance in developing countries	Agricultural Cooperative Development International (ACDI)
Trade association—represent cooperatives on national and international matters	National Council of Farmer Cooperatives (NCFC)

adverse public scrutiny to retard their use. Once these organizations become influential in the market share they handle, they may come under great public scrutiny.

Another problem that may prevent multi-cooperative organizations from being as useful as they could be may result from conflicts among them and the general farm organizations. Cooperatives are owned by farmers with various commodity interests and thus differing agricultural policy objectives. Grain producers, for example, complain if agricultural policy does not provide them with high enough prices to make a sufficient return. On the other hand, cattle feeders want a farm policy that provides cheap livestock feed. Agricultural policy to help cattle producers may be at the expense of the grain producers and vice versa.

Special interest groups have made it difficult for agriculture to become unified in farm policy objectives. The general farm organizations also have differing philosophies with regard to government involvement in agriculture. In many cases, there are conflicting ideas among cooperative organizations, commodity trade groups, and the general farm organizations.

The general farm organizations have been great promoters of agricultural cooperatives. In the beginning, the general farm organizations organized operating cooperatives and worked closely with them. Today, the relationship between the general farm organizations and agricultural cooperatives varies tremendously. Some organizations work closely together, while other organizations compete with each other not only in products but on policy issues. "It's me fighting me," explained one board chairman to his general manager in describing the situation where a general farm organization and an agricultural cooperative were serving the same people.

A few managers were concerned about the need for the various farm organizations to work closer together—"we need to have more meetings on the local level and all spend more time together." Some managers believed that the general farm organizations were slipping in importance; they were not doing much to promote cooperatives; they were getting into areas such as purchasing and marketing where they should not be and, in many cases, they believed the farm organizations did not have the expertise to do an efficient job in these areas.

Managers in the core sample were asked to check which function each multi-cooperative organization (AIC, NCFC, state cooperative councils, the general farm organizations, or other organizations) should be undertaking. All managers indicated cooperative education should be a function of AIC and cooperative legislation should be a function of the National Council (Table 5.4). Most managers believed that public relations was a function for AIC, NCFC, and state cooperative councils. None of the managers believed that cooperative education was a function of the general farm organizations. This is surprising, since the general farm organizations in the past have spent much effort here. One implication, both from the above response and the authors' discussion with regional managers, was that close ties with a general farm organization had few if any advantages for a cooperative. For example, one general manager was very explicit in his feelings that the "social and religious aspects" of a certain general farm organization did not mix very well with the commercial function of his type cooperative.

As far as the conflicts among the different farm organizations, one must ask, What are the alternatives?

TABLE 5.4. Suggested functions of state or national organizations as seen by core sample respondents

	Organization				
Function	AIC	NCFC	State councils	General farm organizations	Other
	Number suggesting particular function*				
Cooperative legislation	1	9	4	3	. . .
General farm legislation	1	4	2	7	1†
Cooperative education	9	3	6	. . .	. . .
Public relations	6	7	7	4	. . .
Social concern	3	2	2	2	1
Farm insurance	. . .	. . .	. . .	4	2‡

* Each respondent checked one or more of the organizations for each function listed.
† National Milk Producers Association.
‡ Includes Mutual Insurance Companies.

One, cooperatives could continue to travel different roads with conflicting interest, some joint effort, and some duplication of effort.

Two, cooperatives could find some common base on which to work together. A manager summarized this idea when he said, "We ought to put a power base together that includes farm organizations, bargaining cooperatives, and operating cooperatives with some semblance of cohesion so we have something to say about the future of agriculture."

Three, cooperatives could have separate organizations, according to functions, with a minimum overlapping of effort but with common goals. This would require division of general farm organizations from ties with operating and bargaining cooperatives. For best results, each farmer would need to be represented by three different types of organizations: (1) his cooperative for purchasing, marketing, and other services, (2) his commodity organizations for certain research and promotional efforts, and (3) his general farm organization for promotion and development of general farm legislation.

ASSESSMENT. Cooperative managers and directors have an opportunity to better carry out cooperative purposes by making greater use of cooperative commonalities. Cooperatives have a number of things in common such as reasons for being, legal sanctions, need for education of members and the general public, and access to the Banks for Cooperatives.

Sharing of commonalities opens up communications among cooperative managers and directors and between cooperatives. By working together, cooperatives can put themselves in a better position to carry out stated purposes.

Several managers did explain commonalities as the ability to work together—to be able to form commercial arrangements together. Since supply cooperatives have worked together to a much greater extent than marketing cooperatives, as one manager expressed it, a "great deal more commonality is in supply cooperatives than marketing cooperatives."

Managers as well as academic people in the selected sample had differing opinions as to whether agricultural cooperatives, historically as a group, have had substantial commonality. Managers had mixed thoughts relative to this same question—both as to the present and the next decade. However, a majority of the academic people believed that at present, agricultural cooperatives do not have substantial commonality, but that over the next decade, commonalities will increase.

Managers did mention some areas where they thought cooperatives did have commonalities: finance, taxation, and legislative matters. Other areas not mentioned that could be considered commonalities are research, education, food and farm policy, and management and director training.

Cooperatives have and are making use of the things they have in common. A number of multi-cooperative organizations provide significant benefits to individual farmer members. Yet there is a question as to why cooperatives do not make more use of their commonalities.

If cooperatives are going to help maintain the family farmer by being a major economic force in the marketplace, they will not be able to accomplish it by each individual cooperative going its own way. Farmers have organized cooperatives for the purpose of enhancing their own economic welfare. Thus each cooperative has this common goal. But the concern of managers and directors is usually not directed to farmers in general but to their cooperative and its owner-members. This behavior is understandable, yet cooperatives will not accomplish some of their common objectives unless some individual cooperatives share some of their inherent advantages. Some specific potential areas—transportation, distribution, marketing, research and development—appear ready for more and improved use of multi-cooperative organizations.

Present multi-cooperative organizations are working fairly well. The interregional cooperatives in supply and fertilizer have been making exceptional growth. The National Council has been gaining in membership strength. Managers were generally pleased with the organization's performance. AIC has continued to improve its education and service role.

One way that cooperatives could work toward helping improve the public understanding of cooperatives is through AIC. To do this would require greater input of cooperative effort into AIC and greater use of the resources and programs that AIC has available. With the proper resources and guidance from cooperatives, AIC could play a much greater role in educating the public about agricultural cooperatives.

Managers in the core sample of cooperatives were asked to check what function could best be done through interregional cooperation (Table 5.5). Energy discovery, basic fertilizer production, and foreign trade were three functions checked by four or more of the respondents. Little interest was shown in coordinating sales, transportation, bargaining information, or general information.

When asked to rank the functions according to what they believed were most important, energy discovery was ranked number one. Basic fertilizer production was a close second. Foreign trade ranked third.

Potential for multi-cooperative organizations exists in marketing, especially livestock and grain exports. Cooperative organizations in private labeling, branding, and other marketing services would be advantageous for farmers. Managers, however, do not think some of these areas would be

TABLE 5.5. Suggested commercially oriented functions that should be done through an interregional cooperative organization as seen by core sample respondents

Function	Organization							
	A	B	C	D	E	F	G	H
Transportation coordination			X		X			
Energy discovery	X	X			X	X	X	X
Foreign trade		X	X		X			X
Basic fertilizer production	X	X			X	X	X	X
Bargaining information system					X			
General information systems				X				
Coordinated sales—domestic								X
—foreign		X						

workable. Said one, "It is not realistic to expect multi-cooperative organizations for branding."

Cooperative leaders need to get farmers to work closer together. Multi-cooperative organizations are aids or strategies that can be used to help obtain cooperatives' reasons for being. Existing and potential organizations must continue to be evaluated by all farm leaders.

Joint Ventures

The joint venture is a specific organizational strategy potentially important for cooperatives wishing to integrate either forward or backward. The authors did not attempt to cover this strategy in great detail but will discuss it in light of research field experience and recent literature on the subject.[44]

CONCEPT AND THEORY. A joint venture is defined as an association between two or more participants organized for the purpose of conducting a specific enterprise, but with the identities of the participants remaining apart from their participation in the venture. Participants share, on an agreed basis, expenses, profits, losses, risks, and some measure of control over the conduct of the operation. A cooperative federation could possibly be viewed as a joint venture of local cooperatives; however, horizontal or vertical coordination seems a more appropriate term for such activity, as discussed above. A contract for the sole purpose of supplying raw materials or purchasing finished goods is not normally thought of as a joint venture.

CURRENT SITUATION. The Hulse-Phillips study identified twenty-two arrangements involving cooperatives in the food marketing area that could be called joint ventures. Fifteen provided marketing services to fruit and vegetable cooperatives (the joint venture between Agway, Curtice-Burns, Inc., and Pro-Fac Cooperatives, Inc., is an example). Seven joint ventures provided services to sugar, livestock, and poultry associations. Services provided by these joint ventures included processing, marketing, sales promotion, transportation, and storage.

The joint venture has taken two forms. In the separate-entity venture, a separate operating unit—usually with a formal organizational structure—is formed to handle the joint venture operation. The Hulse-Phillips study found twelve of these operating in the food marketing industry. In the contractual arrangement venture no separate entity is formed. (Central Carolina Farmers and Gold Kist have formed such a venture.) Ten of these joint ventures were found in the above study.

Joint ventures may be between cooperatives, between cooperatives and noncooperative firms, or between cooperatives and farmers directly. The Hulse-Phillips study found ten ventures between cooperatives, eleven between cooperatives and noncooperative firms, and one between a cooperative and a large farmer.

MOTIVATION. The joint venture is considered here as an organizational strategy primarily to be used with the general strategy of integration and coordination. As such, it has many of the same attractions as does integration through ownership of separate stages in the marketing system.

Possibly the most common reason for a marketing cooperative to participate in a joint venture is the desire to better coordinate marketing efforts. A cooperative skilled in handling production-related problems and skilled in first-stage handling and processing might wish to arrange for sale of its products through another firm with well-established market outlets and management expertise in consumer product marketing and finance. The cooperative would thus be helping assure itself a market for its products through a closer coordination with the food marketing system. Joint ventures in this context are steps to vertically integrate production and/or processing functions with firms operating at higher levels of the marketing system. Typically, firms at these higher levels seek to assure themselves a stable source of raw material.

Other reasons for forming a joint venture include (1) the filling out of a product line by jointly marketing with a company having complementary products; (2) the need for processing facilities, either an existing set of facilities with the skilled management to run owned facilities or the need to fully utilize owned facilities; (3) the need to gain access to research skills and facilities; (4) the need to diversify; (5) the need to tie cooperative management into stock incentives available in the proprietary firm; and (6) the need for logistical improvements. For example, Ag Foods, a federation of cooperatives, is a type of joint venture among six fruit and vegetable marketing cooperatives, formed to provide transportation and warehousing services.

Merger or acquisition is sometimes stated as another reason for joint ventures. The alliance can serve as a "getting acquainted" period, easily broken off if the merger turns out to be unacceptable. This advantage may be of special interest for joint ventures among agricultural cooperatives considering merger. It will usually be much less costly to break off a joint venture than to break off an unsatisfactory merger arrangement.

The joint venture motivation for a noncooperative marketing firm may be the opportunity to assure itself of a dependable supply of raw material. It may also be to round out a product line, utilize a facility, upgrade a product, expand a market, improve a distribution system, or to "spin-off" a low margin seasonal processing facility to a cooperative.

The corporation might also wish to share some of the unique advantages flowing from the cooperative form of business organization. These advantages might include better financial leverage through the cooperative's special sources of credit, untaxed cash flow resulting from deferred patronage dividends, or nonequity surplus.

There is a question, however, whether the joint venture between cooperative and corporation could make any use of cooperatives' fund sources such as the Bank for Cooperatives. Joseph G. Knapp has said, "I can't believe that the resources of the Bank for Cooperatives are going to be made generally available to noncooperative concerns, even through indirect means."[45] Another question arises in regard to antitrust regulations when a joint venture involves a noncooperative firm. Ray A. Goldberg has said, "As a cooperative becomes involved in joint ventures with a corporation, the question arises whether it might lose its Capper-Volstead exemption."[46] Such a loss would probably mean the organization could no longer exist as a marketing cooperative without being in violation of the antitrust laws.

ASSESSMENT. Study emphasis was given to joint ventures between cooperatives and noncooperative firms. The core sample managers were asked in the questionnaire to agree or disagree to two statements: (1) "Your organization could improve its operation by joint ventures with noncooperative firms"; and (2) "Cooperatives will be using more joint ventures with noncooperative firms in the next decade." A majority of the managers agreed to both statements. A majority of the academic personnel also agreed that cooperatives would be making more use of joint ventures with noncooperative firms.

Joint ventures were also discussed during field interviews. Most managers saw the use of joint ventures as a worthwhile alternative to be considered when planning future integrated structures. But use of joint ventures was found to be limited especially between cooperatives and noncooperative firms.

One cooperative manager captured the beliefs of many when he stated, "I don't care who owns it [the joint venture entity] as long as we get the profit." He, for one, was not optimistic on joint ventures with noncooperative firms because of the "complications you run into." The manager did not believe he had any "philosophical hang-ups," but he did think the joint ventures could aggravate problems for cooperatives with the government.

Several managers believed the real potential for the joint venture strategy was for more joint ventures among cooperatives. Such ventures are used only to a limited extent except for the use of cooperative federations. A joint venture among cooperatives is probably free from much of the legal and control problems inherent with a noncooperative joint venture. Cooperative to cooperative joint ventures may be a quite limited means, however, of increasing access to established market channels. The apparent unwillingness of cooperatives having their own effective marketing channels to enter into any form of venture with less fortunate cooperatives enforces this view.[47]

Some special and highly important problems are associated with joint ventures between cooperatives and noncooperative firms. These stem from differences in the basic interests of the two types of businesses. Given these circumstances, it is crucial that the joint venture complement the business of each partner, that there be a mutuality of interest for the functions the venture is meant to perform, that there be a mutuality of respect among partners, and that areas of potential conflict be carefully explored ahead of time. Problems with ventures between cooperatives and noncooperative firms have pointed out that farmers must think carefully about retaining majority control over any processing facilities or marketing channels they do own.

Joint ventures with noncooperative firms seem attractive for cooperatives searching for methods of increasing producer participation in integrated food marketing. However, in today's regulatory climate, cooperatives may wish to consider first opportunities for closer alliances among themselves before aligning with noncooperative firms.

6 Facilitating Strategies

FACILITATING STRATEGIES are necessary to (1) protect, maintain, or improve enabling legislation, (2) develop to the fullest extent the general marketing strategy most applicable, and (3) enhance the organizational strategies that will allow the most efficient method of operations. No classification of such aids and strategies is complete, but five major groupings are (1) acquiring government sanction, (2) using information systems, (3) enhancing domestic demand, (4) expanding foreign trade, and (5) expanding financial sources.

These aids and strategies are sometimes not under the control of the individual cooperative and may require group action. Such action may be joint by several cooperatives, or it may be instigated by the public at large. Most of these arrangements are external to the firm but provide a basic process for the firm to adapt to its environment.

Acquiring Government Sanction

Farmers, through their cooperatives, have had enabling legislation to improve their economic welfare. Concern exists about the abilities of cooperatives to retain certain sanctions given them. Consumers, government regulators, competitors, and others are asking that cooperative legal sanctions be reviewed, amended, or eliminated. Farmers, managers, and other cooperative leaders must consider where they should take a unified stance and where they can afford to be divided on such issues as the Capper-Volstead Act, bargaining, marketing orders, marketing boards, and taxation.

This discussion will cover some of the issues, their importance, cooperative policy with regard to them, and the political climate for each.

CAPPER-VOLSTEAD ACT. The Capper-Volstead Act came into being because of the desire by farmers to join together in associations, without violating antitrust laws, to counteract the power of buyers or handlers. The Capper-Volstead Act exempted the formation of agricultural producer associations with or without capital stock from violating the Sherman Antitrust Act of 1890 and the Clayton Act of 1914.[1]

The act provides enabling legislation for "persons engaged in the production of agricultural products as farmers" to join together in associations and these "associations may have marketing agencies in common." The major provisions of the act are that the associations so formed be "operated for the mutual benefit of the members thereof" while conforming to specified requirements. The requirements are that each member be given no more than one vote because of stock ownership, or the return on stock or membership capital be no more than 8 percent each year. The value of products handled for nonmembers shall not be greater than that handled for members.[2]

A provision was also inserted into the act (Section 2) stating the procedure for regulating farm cooperatives when it is believed that prices are being unduly enhanced by reason of monopolization or restraint of trade in interstate or foreign commerce. The responsibility for enforcing Section 2 was given to the secretary of agriculture.

Importance to Agriculture Cooperatives. The Capper-Volstead Act gives farmers the right to join together in "collectively processing, preparing for market, handling, and marketing in interstate and foreign commerce, such products of persons so engaged."[3] Through joint efforts, or "marketing agencies in common," farmers lessen the competition among themselves, but they can better compete with large-scale buyers of their products. A manager summed it up when he said, "There's no way the growers could survive if they didn't act as a group."

Joseph G. Knapp summarized the importance of agricultural cooperatives when he concluded that

> the cooperative form of enterprise . . . need not and should not ever completely supplant other forms of business. The cooperative form of enterprise both complements and supplements the services performed by other forms of private business, thus giving our system even greater flexibility and strength.
>
> By providing a self-help mechanism through which people and business firms can serve themselves according to their needs, the cooperative can also democratize and decentralize parts of our economic life, provide pacesetting competition for other forms of business, and give the individual a sense of belonging. It can act as a balance wheel—or a safety valve—in our economy by providing an alternative type of business organization within the free-enterprise system that we value so highly.[4]

Although the Capper-Volstead Act does allow farmers to join together, it does not give them complete immunity from the antitrust laws. A cooperative like any other business enterprise cannot abuse its place in the market.

Undue price enhancement as previously mentioned is prohibited by Section 2 of the Capper-Volstead Act. Other predatory or illegal practices are also prohibited by the same laws governing other business enterprises.

The specific provision in the Capper-Volstead Act that allows cooperative associations to have marketing agencies in common is an important aspect of the act. This arrangement permits autonomy of local associations and closer control by farmer members. No counterpart to this provision exists in the antitrust laws that applies to noncooperative businesses. Many people believe it is a distinct advantage for cooperatives, as well as being in the spirit of farmer cooperation. However, none of the cooperative managers in the core sample believed that their organizations had any important advantages over their competitors due to the legal support given them by the Capper-Volstead Act.

Cooperative managers were found to be perturbed, however, about attacks by certain public officials and others on cooperatives and the Capper-Volstead Act. They were worried that these pressures might result in amendment or even repeal of the act.

The authors asked managers what they would do if the Capper-Volstead Act were lost. Managers of a few organizations said that they had given thought to converting their cooperative to some type of corporate form of business as a means of surviving. One manager believed that without the Capper-Volstead Act "there would be a real problem trying to restructure in such a way that we could pass through the profits to the farmer." Others believed that if the act were lost "it would be the consumers who would be hurt most in terms of numbers of people," because competitors would gain more control, resulting in higher food prices.

Other managers were concerned about the "unknown effects" if the act were brought up for review by Congress. They believed that any action to change the act would greatly impair the family farmer and competition, and hurt the United States economy.

All managers and academic people agreed (50 percent strongly agreed) with the following statement: "The Capper-Volstead Act is important if agricultural cooperatives as a group are to meet their current objectives." With only one exception, the respondents believed that the act would continue to be needed for this purpose in the future.

Legislative Policy of Cooperatives. Cooperative managers were very uncertain about what could or would happen to the Capper-Volstead Act over the next five years. As one manager put it, "winds are blowing too strong" with regard to the act. Many managers believed that there was a good chance that within the next five years the act would be evaluated by Congress.

The response from the academic people was quite different. They believed the chances of the act being evaluated by Congress in the next five years was much lower. Most managers in the core sample disagreed that Congress should make a thorough evaluation of the Capper-Volstead Act. The academic people were evenly divided on the idea.

Cooperative managers were asked where they thought changes in the

President Harding signing the Capper-Volstead Act on February 18, 1922. The act gave farmers the right to form marketing cooperatives without being in violation of antitrust laws.

158

Capper-Volstead Act might occur. Of the possible changes that might occur, two were most feared by managers. One was the transfer of Section 2 out of the USDA and into the Justice Department. A manager stated the problem this way: "Justice has absolutely no feel for agricultural cooperatives." Another was managers' concern about deleting the "marketing agencies in common" provision. As one manager put it, "The ability to operate on a federated or interregional basis is crucial."

Some of the current policy questions regarding the Capper-Volstead Act and operation of agricultural cooperatives are: Who is included in the definition of a "producer"? What constitutes "democratic control"? What is "undue price enhancement"? What ways can cooperatives combine efforts? What is a permissible market share? What is the required method for handling products for nonmembers versus members?

Many cooperatives belong to the National Council of Farmer Cooperatives and make their wishes and desires known about legislative policy through the council. The scope of the council's activities covers a program of action to represent the members before all branches of government—executive, legislative, and judicial—in such a way as to influence government statutes and regulations in accordance with legislative policy decisions approved by the membership. Currently, the council is fighting vigorously to protect the Capper-Volstead Act.

Some of the large regional cooperatives also finance individual lobbyists and consultants. Managers of these organizations believed that it was not possible for the council to take care of all member needs. The council must concern itself with the majority interest, and in some cases this limits their activity for a specific cooperative.

Managers expressed a serious need for agricultural cooperatives to educate the public and legislators to what the cooperatives are all about. Several managers thought the area of highest priority was in articulating the role and functions of agricultural cooperatives to the legislative branches.

Only a few organizations were attempting in any major way to keep their farmer members informed as to any anti–Capper-Volstead action. Several managers believed that any serious action against the act would take time and farmers could be informed. "Serious consideration to weaken the act," explained one manager, "would bring great action on the part of farmers."

Just how well informed should members of agricultural cooperatives be with regard to the Capper-Volstead Act and other policy issues? Many have been concerned that cooperatives need to do a better job of educating their members on policy issues. Only two core sample managers answered affirmatively that "our membership is fully aware of the serious nature of the current policy issues endangering the legal support of agricultural cooperatives."

Political Climate. The Capper-Volstead Act is under attack. Possibilities exist that the act will be reviewed and changed. Several factors are creating this critical climate.

First, inflation—especially evidenced by higher food prices and higher than normal unemployment—has prompted consumers and policymakers to question antitrust exemptions and regulations. Bigness, profits, and special legal treatment become the prime suspects for the cause of high food prices. Consequently, cooperative officials are continually defending their cooperatives' reason for being.

Second, actions by a few dairy cooperatives have attracted much attention over the past decade. This attention was due to a few dairy cooperatives' illegal contributions to the 1972 reelection campaign of former President Richard M. Nixon and several federal antitrust suits alleging price fixing and coercive marketing practices.[5] These actions have hurt the image of cooperatives in general.

Third, studies by various agencies and remarks by some public officials have focused increased attention on cooperatives, the Capper-Volstead Act, and the marketing order system.

In March 1974 the Agribusiness Accountability Project, in Washington, D.C., issued a report entitled *Who's Minding the Co-op?*[6] The report claimed that farmers were losing control of their cooperatives and that the bigger cooperatives become, the less sensitive they are to their members' needs.

Also in 1974 a Republican Task Force on Antitrust and Monopoly Problems, chaired by Congressman J. Henry Heinz III of Pennsylvania, was organized to examine the competitive conditions in the food and agriculture, energy, and communications industries. In October 1974 a staff report on food and agriculture, which was to be kept confidential, because as Heinz states, "We felt strongly there were some people we had not heard from and the conclusions expressed in it were at this point premature," was leaked to the Associated Press.

The report recommended "an amendment to prevent food-processing corporations from gaining antitrust exemptions by forming farmer-style cooperatives"; "a change in antitrust laws to allow the Justice Department to get court orders to prevent mergers of farmer co-ops into giants"; "an end to bloc voting in establishing federally regulated markets for milk, fruit and vegetables"; and other changes. The report also stated that some cooperatives were "approaching monopoly positions" and some were using undemocratic methods.[7]

In January 1975 Keith I. Clearwaters from the Department of Justice in a speech to the American Farm Bureau stated that the Justice Department suggested to the House Judiciary Committe in 1974 "the need for Congressional reevaluation of antitrust immunity for cooperatives to determine, among other things, the degree to which the activities of cooperatives enhance food prices, to determine whether some size limitation should apply to cooperatives so that they do not dominate national or regional food markets, and to determine the effect of vertical integration by members of cooperatives into processing, as well as the size and functions of individual members."[8]

In February 1975 the President's Council of Economic Advisers told

Congress that "some of the largest cooperatives may have gone beyond the original intent of the legislation, and with the aid of agricultural marketing orders may have been able to maintain certain commodity prices above competitive levels."[9]

The Justice Department believed that some cooperatives were charging above competitive prices, and since there was no apparent action on the part of the secretary of agriculture under Section 2 of the Capper-Volstead Act, they urged Congressman Robert McClory of Illinois to introduce legislation in April 1975 that would transfer jurisdiction of Section 2 from the Agriculture Department to the Justice Department.

In September 1975 the Federal Trade Commission released a *Staff Report on Agricultural Cooperatives.* The "study was initiated to examine the intended scope and present status of the agricultural cooperative antitrust law exemption." It was "concluded that the status of agricultural cooperatives under the antitrust laws is roughly on a par with that of corporations." The study also briefly reviewed the economic impact of cooperatives and made an analysis of federal and state "marketing order" programs.[10]

On April 5, 1976, the USDA began a review of dairy cooperatives' pricing practices in federal order markets. The purpose was "to ascertain whether there is reason to believe that cooperatives have unduly enhanced milk prices in violation of the Capper-Volstead Act."[11]

The situation that brought about the USDA inquiry was the sharp increase in the number of markets with and size of over-order payments in late 1974 and early 1975.

The USDA concluded, "Considering the national average picture, there is no evidence that cooperative actions in obtaining over-order payments in 1974–75 resulted in prices that were too high for economic conditions. The economic evidence confirms the contention of the cooperatives that the sharp increase in over-order payments in this period reflected their efforts to offset the sharp drop in Federal order Class I prices at a time of a severe cost-price squeeze on their members, the producers."[12]

On January 17, 1977, a report was issued by the Antitrust Division of the Justice Department on milk marketing. The purpose of the report was to assess the competitive impact of federal milk market orders, the conduct of dairy cooperatives, and the interaction of the two.

The authors briefly reviewed the legislative history of the Capper-Volstead Act. From observations of milk marketing in the 1920s and 1930s to milk marketing today, the authors raised some issues about the continuing need for federal market orders and the proper scope of the act.[13]

In a USDA response to the Justice Department report, the current scope of the act is defended. The USDA stated that "the law [the Capper-Volstead Act] as written has been flexible enough to permit cooperatives to operate well in a changing economy, but cooperatives have not been exempt from prohibition against anticompetitive and unfair trade practices. It should be noted also that officers, directors, or agents of a cooperative like those of any business corporation may be and have been prosecuted if found responsible for an antitrust violation.[14]

On December 1, 1977, a National Commission for the Review of Antitrust Laws and Procedures was established by the president. The functions of the commission were to study and make recommendations on subjects involving (1) ways to expedite the resolution of complex antitrust cases and (2) "the desirability of retaining the various exemptions and immunities from the antitrust laws, including exemptions for regulated industries."[15] Two major agricultural issues under study by the commission were the Capper-Volstead Act and the marketing order system.

At a commission public hearing on July 27, 1978, Secretary of Agriculture Bob Bergland stated:

> My own view, I believe well supported by history, experience and research, is that the Capper-Volstead Act and our marketing order system are in no need of statutory modification.
>
> Actions to modify these agricultural provisions may be intended to increase competition, but they may in fact weaken competition. The buyers' side of the agricultural product markets has gained tremendous strength because of buyers' size. But the producer's side is still made up of individual farmers. Their only realistic hope of some equity in the market is effective cooperation.[16]

The General Accounting Office has recently begun a study of agricultural cooperatives. The study was self-initiated to address current cooperative issues, in particular to identify (1) the role of cooperatives in today's agricultural environment and their importance to the family farmer; (2) the farmers served by cooperatives; (3) the need for better monitoring of cooperative activities; (4) the need for changes in cooperative laws; and (5) the strengths and weaknesses in the USDA's cooperative research and technical assistance programs.

A report is expected to be issued to Congress in April 1979.

Managers were worried about the political climate. One manager said, "It seems the issue goes up and down, being up right now because of the high food prices." "There seems to be some misconception," explained another manager, "that cooperatives should represent [handle] strictly raw products and should not infringe on marketing operations."

Many managers believe that both Congress and the Justice Department are the biggest advocates for changes in the Capper-Volstead Act. A legal counselor in one cooperative said, "The whole reason for growers getting together is to enhance their position and this was the thrust of the Capper-Volstead Act. However, that position all translates to price, and if price is unduly enhanced, this turns into a monopoly. The key is predatory practices—in other words, undue control by a given sector of a given market. The undue part has really never been tested."

Due to changes in the political climate, mention was also made of the need for more lawyers trained in the area of agricultural cooperatives. The problem is that the "average lawyer never runs into special cooperative problems." It was suggested that regional cooperatives need to build their own expertise because they are the only ones big enough to do it. Local cooperatives could also benefit from the services of a lawyer if they were offered by the regional cooperative.

Changes in the economic environment—such as a lower rate of inflation and unemployment—may bring changes in the political climate affecting cooperatives and the Capper-Volstead Act. Farmers and their cooperatives, however, cannot depend on such changes alone. Managers and all members need to keep themselves informed as to what is happening politically. Managers and directors need to know what actions may result in legal entanglements so that adjustments can be made to avoid actions that give cooperatives a bad public image.

BARGAINING. Bargaining was previously discussed as an alternative general marketing strategy for groups of agricultural producers. This section looks briefly at bargaining in light of government sanction needed for its success. Some views of managers and professors are also reported.

Importance to Agricultural Cooperatives. If farmers are to use the bargaining strategy aggressively, government sanctions will probably need to be strengthened. Growing concentration in the food industry makes the power struggle more intense and negotiation of terms of sale less automatic. Production and sale of agricultural products under contracts place producers at some disadvantage, and they believe they deserve government sanction.

Bargaining, if widespread among cooperatives or if used as a national price and income program focus, would develop wide public interest. Hammering out the appropriate public sanctions would then be involved and would be a difficult matter.

Bargaining as a concept underlies the reason for being of cooperatives. Power of individual farmers had to be welded into group power if bargaining was to be equitable with larger buyer interest. Also, the information base had to be upgraded for farmers. These and other reasons for special cooperative treatment generally are intertwined with the more aggressive posture of modern cooperative bargaining, and public sanctions must be forthcoming to facilitate this thrust.

Today, farmers face situations similar to the historical ones. Also, buyers are larger and fewer. Direct buying and increased contract production have diminished the open market system—the price-discovery mechanism for many commodities. Some buyers are highly vertically integrated. Other buyers are divisions or subsidiaries of huge corporations. This concentration gives them excessive market power in dealing with the individual farmer.

Effective collective bargaining does provide farmers some countervailing power as well as a price-establishing mechanism.

As a strategy, bargaining may be used widely as an aid to the survival of the individual family farmer. For this, and other reasons, collective bargaining by agricultural producers will require specific government sanctions.

Legislative Policy of Cooperatives. No common bargaining policy exists among cooperatives as far as bargaining is concerned. Many managers of operating cooperatives have little feel for agricultural bargaining. The general feeling among these managers was that farmers should do more on their own. They should vertically integrate and get into processing, thereby owning their own operations.

In California the story is different. Managers of operating cooperatives also see a need for bargaining associations to represent them in establishing price and terms of trade. Many of them believe that farmers not only should integrate vertically but should also have an effective bargaining association.

Bargaining cooperatives and other organizations interested in agricultural bargaining have been seeking legislation to give them more and improved bargaining rights. This study, however, found no general support among operating cooperatives for substantially strengthened bargaining legislation.

The National Council of Farmer Cooperatives has been trying to obtain more bargaining cooperatives as members. Currently, only a few bargaining cooperatives belong to the council. The reason for this situation may be due to the fact that (1) many bargaining cooperatives operate on a very thin budget and/or (2) bargaining cooperative leaders do not quite realize what the council does and can do for them. The same enabling legislation—the Capper-Volstead Act that allows farmers to join together in operating cooperatives without being in violation of antitrust laws—also allows farmers the same rights to join together in bargaining cooperatives.

The council's policy resolutions recognize the importance of bargaining programs where desired by farmers.

> The National Council . . . strongly encourages bargaining associations to take advantage of every opportunity to coordinate their activities with operating [full marketing] cooperatives in ways which will maximize benefits to farmer members of both groups. The National Council supports legislation designed:
> 1. To authorize associations of producers of agricultural products to bargain singly and jointly for price and other terms of trade involved in the more effective selling of their products, and to enable bargaining and operating cooperatives to work more effectively for the benefit of producers.
> 2. To authorize cooperative associations to initiate and utilize federal or state marketing orders or agreements and to employ common bargaining or marketing agencies.
> 3. To protect cooperatives from actions by buyers of farm products which in any way may hinder, obstruct or interfere with the formation and operation of such farmer-controlled cooperative associations.
> 4. *To require bargaining in good faith between bargaining associations and proprietary handlers and processors.*[17]

Fewer than one-half of the academic people thought that agricultural cooperatives as a group would be able to bargain effectively under current laws over the next five years. One-third of the core sample managers thought this issue was not relevant to them. Another one-third were undecided whether current laws were effective enough for bargaining by agricultural cooperatives.

Political Climate. At the federal level, new legislation to improve bargaining appears to be at a standstill. Bargaining associations operate under the same enabling legislation (Capper-Volstead Act) as other marketing cooperatives. Thus the political climate surrounding collective bargaining in agriculture is currently fairly poor.

The Agricultural Fair Practices Act in 1967 (S-109) was the last federal law passed concerning agricultural bargaining. The law was enacted to protect farmers and cooperatives from discriminatory practices of buyers or handlers. A number of practices were defined as unlawful by the act.[18]

Since passage of S-109, several other bills (Mondale Bill, Sisk Bill, Hart Bill) were proposed to strengthen collective bargaining in agriculture. Attempts were made at establishing some type of national agricultural bargaining board, providing standards for accrediting association of producers, establishing as an unfair trade practice the refusal of buyers and handlers to negotiate prices and other terms of sale with agricultural bargaining associations, and the requiring of handlers to bargain in good faith with accredited bargaining associations.[19]

Agricultural bargaining is undergoing many challenges. *First,* some people are questioning whether bargaining groups are violating the antitrust laws. Concern is that the intent of the Capper-Volstead Act was not for the purpose of providing farmers the right to collective bargaining. *Second,* federal and state legislation needed to improve the agricultural bargaining process is moving slowly. New legislation passed in Michigan in 1973 was thought to be a benchmark for other states. It gave farmer bargaining associations the right to exclusive bargaining with handlers and processors. The act has been the target of several lawsuits. One lawsuit filed in 1974 has been carried to the state supreme court. A supreme court decision is expected sometime during 1978.[20] *Third,* farmers lack support for bargaining from other interests. Most cooperative managers and farmers not involved with bargaining are failing to support farmers who think they need a price establishment mechanism.

One debate focuses on whether bargaining should be mandatory or voluntary for producers of a commodity. Most bargaining groups operate where membership is on a voluntary basis. The problem with this method is that producers come and go. When the product is in short supply and prices are high, producers believe they can get a better price outside the association. If the product is abundant and prices are low, producers want to market their product through the association. "Keep me long in supply and I wouldn't have any problems with growers," stated one California bargaining association manager.

Bargaining associations have a similar problem with handlers and processors. As long as bargaining in good faith is not mandatory, processors may go where they believe they can purchase the raw product at the lowest price. When an abundant supply exists, processors are more likely to go directly to the growers to purchase their raw product than when supplies are short. Thus the bargaining association has the growers when it does not have the processors, and vice versa.

Another important issue has been whether bargaining cooperatives should take title to the product. Some observers believe that if a bargaining cooperative took title to the product, bargaining would be more responsive to members and processors, would market greater quantity and quality of products, and would better fulfill the intent of the Capper-Volstead Act.

The Capper-Volstead Act does not explicitly provide that producers may form associations for bargaining purposes with antitrust immunity. However, in a case in the Ninth Circuit Court of Appeals on April 11, 1974, involving two bargaining associations and processors, the court ruled that, according to the definition of marketing, the bargaining associations' activities were marketing functions.[21] The court in the case, *Treasure Valley Potato Bargaining Association* v. Ore-Ida Foods, Inc., 497F. 2nd 203 (9th Cir. 1974), established that "farmer cooperative bargaining associations qualify for the anti-trust exemptions, for both Section 1 of the Capper-Volstead Act and Section 6 of the Clayton Act."[22]

In deciding the above case, the court of appeals referred to the 1953 edition of *Websters' New Collegiate Dictionary* for the definition of "marketing." Marketing is defined as "the aggregate of functions involved in transferring title and in moving goods from producer to consumer, including among others buying, selling, storing, transporting, standardizing, financing, risk bearing, and supplying market information." According to this definition, bargaining associations are performing marketing functions whether they take title to the product or not. However, some people do not accept this as the final resolution of the issue.[23]

Future legislation involving agricultural bargaining at the national level will depend on what happens to raw product prices and to food prices, the changing structure of the food industry, and the results of state bargaining legislation passed in Michigan, Minnesota, Maine, and Ohio.

Effective bargaining needs substantial public sanctions. Cooperatives and supporting agencies (i.e., ESCS, colleges, farm groups, etc.) should know what producers want and need in public policy so that they will be prepared to comment if momentum in bargaining legislation is stimulated by changing economic conditions.

MARKETING ORDERS. Marketing orders are often referred to as farmers' self-help programs. The intent is to provide a means whereby farmers can exercise some degree of control of the quality and quantity of products that go to market so that their income will be more stable and the quantity and quality going to consumers would be more stable and uniform.[24]

Because of the self-help concept, many people associate marketing orders with cooperatives. Actually, a marketing order is a distinct program. It can be initiated and carried out with or without a cooperative. However, cooperatives are the logical group to request an order. Cooperatives are usually the catalyst to the organization and operation of orders.

By stabilizing and perhaps enhancing the income of producers, marketing orders accomplish one of the reasons for having cooperatives. In some ways, marketing orders strengthen the marketing or bargaining ability of cooperatives. Thus it is important that cooperatives help maintain the necessary government sanction that facilitates orderly marketing of their patrons' products.

Marketing orders originated under the Agricultural Marketing Act of 1937 as an attempt to stabilize agricultural production and farm prices. The basic concept was for restoration and maintenance of orderly marketing.

Any interested persons can request a federal marketing order for a market for any of the commodities covered by the act. Once a marketing order is proposed, the USDA makes a preliminary investigation. If the proposal is found to effectuate the declared policy of the Agricultural Marketing Act, public hearings are held on the proposal. All interested parties have the opportunity to testify for or against the proposed order.

After the hearings, a preliminary statement and recommended order are published. An additional period of time is given for interested parties to file exceptions to the proposed order. The secretary of agriculture will issue the marketing order after he reviews the exceptions to the recommended order, the hearing record, briefs, and arguments.

The order becomes effective following approval through a referendum by two-thirds of producers (by number or volume of production).[25] Cooperatives may approve or disapprove the order for all members whose product is marketed under the order. All handlers and producers in the marketing area are bound by the order once it becomes established.[26]

The secretary of agriculture has the responsibility of administering federal marketing orders. Usually, a board of farmers will provide advice for the secretary. Costs of administering programs under marketing orders are collected from handlers and farmers on a per unit basis. No tax funds are involved.

Marketing order programs may be sanctioned under federal and state legislation. Most milk, fruit, and vegetable marketing orders are operated under federal enabling legislation. It is possible, however, to have both state and federal marketing orders operating in one state. In New York State, for example, milk producers market milk under both state and federal marketing order programs.

Forty-five fruit, vegetable, and tree nut federal marketing order programs were in effect during fiscal year 1977 in the United States. Fifteen orders were operating in California. Eight of these orders included other areas outside California. Federal marketing orders covering fruit include about one-fourth of the supply within the regulated areas. And orders covering vegetables include about 40 percent of the supply within regulated areas. Federal orders covering speciality crops generally cover the total supply within the regulated areas.

Approximately one-fifth of the total United States fruit supply, less than 15 percent of the total United States vegetables, and nearly 100 percent of the specialty crops—such as dried fruit, nuts, and hops—subject to regulation were produced under marketing order programs.[27]

Advertising and promotion, research, mandatory labeling and inspection, disease control, and similar activities are the main thrust of these orders.

In the fruit and vegetable industry, three general types of programs exist: grade, size, and quality limitations; rate-of-flow regulations; and volume controls. The volume control regulation may include one or more of the following:

1. Market allocation—sets the maximum amount to be marketed in the primary (domestic) market.

2. Reserve pool—allows storage of a commodity in a reserve pool for marketing at a later date.

3. Producer allotments—authorizes the prorating of marketable quantities among producers.[28]

In the dairy industry, forty-seven marketing order programs were currently in operation as of January 1, 1978. "federal milk marketing orders set *minimum* prices for raw fluid grade milk which must be paid by processors to dairy farmers [usually through farmer cooperatives]. All other provisions of the orders are subsidiary to the minimum price. No federal order limits the quantity of milk produced or marketed."[29] In 1977, 66 percent of all milk delivered to all plants and dealers in the United States was covered under federal milk orders.[30]

Importance to Agricultural Cooperatives. About half the cooperatives in this study were affected by marketing order programs. In the core sample, only three cooperatives stated they were affected by state and federal marketing orders. Most managers, in cooperatives that were not affected by marketing orders, had little knowledge of their use.

Cooperative managers (especially in California) did voice some concern about the attack by consumers on marketing orders, consumer pressure to have more representation on the marketing order boards, and use of volume or supply control regulations.

Managers were very explicit in explaining the importance of marketing orders in stabilization of quantity and price, setting uniform grade standards, research, promotion, and market information. Marketing orders "prevent gigantic surplus situations from wrecking the market," stated a manager. However, he believed that in the long run supply has to be balanced with demand to prevent market disaster. Some managers believed that consumers as well as Congress did not understand marketing orders and what they really represent. "Their [consumers] whole feeling," stated one manager, "is that if you take away marketing orders you're going to have lower food prices."

Many of the managers in California were unhappy with the pressure for greater consumer representation on the boards. They feared consumer representatives would not understand their problems. However, a few managers believed that having consumers on the boards would be a good way, possibly the only way, for producers ever to get consumers to understand marketing orders and the problems of producers.

Managers believed the most important issue was public resistance to the use of supply controls in marketing orders. As one manager put it, "Supply control is the pivotal issue. Most people would accept marketing orders on the basis of research, producer finance, promotion, and advertising. However, when you get into supply control, then the issue is very open." Some managers believed the whole idea of using supply control in certain marketing orders was a dead issue. Even growers had expressed dissatisfaction with the use of supply control. However, concern was expressed that the attention given supply control was being used to sidetrack the real benefits producers and consumers receive from marketing orders.

It is often expressed that marketing orders are beneficial only to cooperatives. Not surprisingly, questionnaire responses from managers who were affected by marketing orders all disagreed. They believe others such as processors, consumers, or distributors may also benefit.

Nonetheless, some experiences have shown that marketing orders do provide a framework for cooperatives to preserve competition and to maintain orderliness in the market. One experience was reported by Herbert L. Forest in discussing the reasons for one cooperative's request for a marketing order in a territory where an order had previously been terminated.

> The reason for this turnaround was the development of disorderly marketing conditions.
> —Producers had been cut off;
> —There was no effective program for pricing milk according to use;
> —There was no impartial audit of handlers' records to verify the payment of applicable prices;
> —There was no system for verifying the accuracy of weights and butterfat content of milk purchased; and
> —There was no method of providing uniform returns to all producers supplying the market.[31]

A number of marketing experts have concluded that marketing orders can strengthen bargaining and marketing cooperatives.[32] However, the importance of cooperatives in helping carry out marketing order programs appears to be downplayed. A cooperative can sponsor and ensure continuation of the order. Olan D. Forker and Brenda A. Anderson stated that "historical evidence suggests that an active group within the industry, willing to play an important role in the formulation and operation of the order, provides the basis for potential success and continuation of the order."[33]

Cooperative managers, in general, were not united in their thoughts about greater use of marketing orders. A lack of consensus existed as to whether cooperatives as a group should move to explore greater use of federal or state marketing order applications in their programs. But they foresee federal marketing orders as fulfilling more of a role for orderly marketing than state marketing orders.

Most respondents, when asked about ways in which marketing orders could become effective in cooperatives' programs, were in agreement that greater use be made of volume control through market allocation but not through production control at the farm or ranch. A majority of the respondents believed orders would be effective for product promotion and development programs. They did not see the addition of consumers or laborers to the boards as making the orders more effective for cooperative programs.

A majority of the academic people believed that cooperatives would not make greater use of marketing orders in the next decade. This does raise some questions about the future of marketing orders. Will future producers without formal cooperative effort be able to initiate the process in obtaining a marketing order or to maintain the government sanction needed for successful orderly marketing? Should cooperatives, on behalf of farmers in general, place greater emphasis on the use of marketing order programs?

Legislative Policy of Cooperatives. A policy resolution adopted by the National Council of Farmer Cooperatives sums up cooperative policy with regard to marketing orders. It states that cooperatives favor

> the development of federal enabling legislation to make possible the formulation of producer marketing orders to develop intra- and interstate and national marketing programs . . . such orders should enable producers to develop marketing programs *as needed to ensure long-term reliable food supplies for consumers* in the following categories:
> 1. Grade, size and quality standards
> 2. Container standards
> 3. Promotion-advertising
> 4. Market service-merchandising
> 5. Research
>   a. market statistics
>   b. economic price analysis
>   c. production problems
>   d. handling and engineering problems
>   e. new product development
> 6. Third party grading
> 7. Surplus control[34]

Political Climate. The political climate surrounding marketing orders has been similar to that surrounding the Capper-Volstead Act. Justification of milk marketing orders has been questioned with intensity because of their direct interference in the pricing mechanism and the large share of producer milk handled by dairy cooperatives.

Consumer groups and some legislators are questioning the over-order premium prices charged by cooperatives. They argue that cooperatives are able to charge the higher price from handlers because of the government sanction provided by marketing orders.

The concern over the effects of marketing orders gained impetus from the increasing rate of inflation and the rise in food prices in 1972. Price increases were at their peak between 1973 and 1974. The consumer index from 1973 to 1974 increased almost 11 percent. Due to these prices, before a joint session of Congress in October 1974, the president's address included this fact: "Agricultural marketing orders and other federal regulations are being reviewed to eliminate or modify those responsible for inflated prices."

In November 1974, because of the concern over inflation, an interagency Task Force was created to investigate price impacts due to federal market order programs. The Task Force concluded, with respect to milk marketing, "that federal milk marketing orders have some significant effect, over time, on under-girding over-order prices." It also stated that on balance, the classified milk pricing system is not inflationary.[35]

The fruit and vegetable marketing orders were evaluated according to type—grade, size, and quality limitations; rate-of-flow regulations; and volume controls. In general, the Task Force concluded that for many of the commodities, marketing orders did not have a significant price impact. However, potential price enhancement capabilities were found to exist for some commodities.

The *Economic Report of the President,* February 1975, included a chapter on "Government Regulation" questioning the "efficacy of regulation." Reference was made to the Capper-Volstead Act and agricultural marketing orders as possible aids to enhance prices above competitive levels.[36]

The Federal Trade Commission in September 1975 issued a report dealing with "an analysis of the agricultural cooperative antitrust exemption, agricultural 'marketing order' programs, and several collateral issues relating thereto."[37]

Marketing orders, it was concluded in the report, "have contributed to the ability of agricultural cooperatives to gain and maintain market power." Cooperatives have also sought for marketing orders to restrict supply.[38] It was also concluded that "marketing orders do have potential and actual effects counter to consumers' interests." Misallocation of resources was stated as being the result.

The USDA responded to the FTC Staff Report, claiming the report was "misdirected" and had taken "incorrect approaches to farmer cooperatives."[39]

With respect to marketing orders, the USDA stated that

> The report presents a particularly inaccurate and distorted picture of marketing orders, since it deals only with what are admitted to be the "most clearly restrictive aspects" of such programs. Further, these are portrayed in terms of what might happen if fully exploited to industry's advantage, without recognition of the complete control exercised by this Department in the issuance of orders and regulations.[40]

The USDA also questioned the report's accuracy in several areas.

Aileen Gorman, Executive Director, National Consumers Congress, stated in November 1975 that her organization strongly urged the USDA to support the following change: "That the Federal Market Order system be retooled to reflect the fact that we are moving toward a single class of milk. This would require amending the Agriculture Marketing Agreement Acts and mandating the Secretary of Agriculture to develop new pricing mechanisms which reflect modern market conditions."[41]

In December 1975, the Community Nutrition Institute sponsored a Conference on Milk Pricing and the Market System. Issues such as the large dairy cooperatives, multiple pricing under the federal marketing orders, nonfunctional over-order premium prices, the social cost of milk marketing orders, and government regulation were all discussed. The major impetus of the conference was for reform of the federal milk marketing orders.

At the 20th National Conference of Bargaining and Marketing Cooperatives, in January 1976, Floyd F. Hedlund, Director of the Fruit and Vegetable Division of the Agricultural Marketing Service (AMS) in the USDA, referred to the attention marketing orders have obtained from various agencies and consumer groups. His response was that "marketing orders must be able to withstand the scrutiny of thorough analysis and review. We cannot simply complain that someone is invading our turf. On the other hand, we must insist that studies and analyses are indeed competent, objective, and thorough."[42]

On March 5, 1976, the secretary of agriculture established an Advisory Committee on Regulatory Programs. The function of the committee was to review selected regulatory programs and recommend either methods of improvement or their elimination. Members on the committee represented the food industry, the farming community, consumer interests, and the university sector.[43]

The committee selected two subject areas to review: (1) marketing orders on milk, fruits, and vegetables, and (2) regulatory activities of the meat and poultry inspection programs.

On July 22, 1977, the committee, with respect to marketing orders, made the following recommendations:

> 1. That order committees provide for public representation. . . .
>
> 2. That adequate notice of all meetings shall be publicized in the affected production areas.
>
> 3. That a study be initiated of alternative methods for establishing minimum prices under the milk order system with the objective of creating a more stable and predictable market situation, and exploring particularly the advantages/disadvantages of establishing one price to cover all expenses.
>
> 4. That a system be developed to assure that a wide variety of viewpoints will be represented at public hearings (on milk marketing orders) by encouraging consumers to testify and providing notification of hearings at the lowest level possible.

5. That a solid research program be initiated on the cost of milk regulation in relation to the benefits therefrom.[44]

In April 1976 the National Milk Producers Federation held a Symposium on Cooperatives, Milk Marketing, and Regulations. The symposium was held because of the controversy and questions concerning cooperatives and federal milk marketing orders. The federation's aim was "to [help] lay the foundation for an improved understanding of the issues and problems involved."[45]

In June 1976 an article appeared in the *Baltimore Sun* claiming that critics had found one big factor in the cause of food price rises—the marketing order program. The story was based on a government study completed by the General Accounting Office (GAO). The GAO claimed the higher prices were due to producer committees imposing quality control regulations and keeping part of the crops from the fresh market. The GAO estimated the inflationary impact of marketing orders on potatoes and onions at about $7 million for 1974.[46]

On January 17, 1977, the Antitrust Division of the Justice Department issued a report on milk marketing. The report was prepared by a created Antitrust Immunities Task Force as part of a broad administration effort to evaluate and reform government regulation of business.[47]

Even though a number of recent studies have dealt with cooperatives and milk marketing, the Task Force reported that "few of these studies have attempted to examine the entire context of the subject from an independent and pro-competitive viewpoint."[48]

The authors concluded that cooperatives to some extent have been able to control the movement of some milk in an order through mergers, federations, and other collective activity. And in various regional markets, cooperatives have been able to achieve monopolistic market power. "In addition," stated the authors, "the federal order system, whose provisions have on occasion become predatory tools, has also been instrumental in preserving the market power of dairy cooperatives."[49]

The authors also concluded that the order system has caused or contributed to chronic oversupplies of grade A milk; inefficient, inequitable, and regressive attempts to raise producer incomes; accentuated fluctuations in prices received by farmers; artificial localization of supply areas; and monopolistic pricing by dairy cooperatives.[50] The authors suggested some possible remedies to the problems they identified.

The USDA response to the report took exception to a number of views and conclusions: "The Department of Justice authors presented a massive amount of material to support their conclusions. However, they marshalled their 'facts' to support conclusions a well-trained researcher would consider to be only testable hypotheses. The report is an adversary brief, setting forth alleged undesirable aspects of milk regulations."[51]

The USDA listed some serious "misimpressions" that could arise from accepting the Justice report at face value and also commented on some of the major issues the report raised such as cooperatives and the Capper-Volstead Act, cooperatives and marketing order operation, and costs of milk regulation. Further comments were also included on specific statements included in the report.

In November 1977 a condensed and slightly edited version of the Department of Justice report on milk marketing was published by the American Enterprise Institute, entitled *Federal Milk Marketing Orders and Price Supports.*[52]

On December 1, 1977, the president created a National Commission for the Review of Antitrust Laws and Procedures. One area of concern to the commission was the federal marketing order system.

The above examples are not comprehensive but they do provide a picture of the climate regarding marketing orders. Their use and function are under serious questioning by a variety of influential groups. Thus much uncertainty exists with regard to the political climate surrounding marketing orders and their future. It appears that farmers need continuously to evaluate the possible effects on their cooperatives of changes in marketing orders due to changes in legislation.

MARKETING BOARDS. Marketing boards can be used to perform a number of functions. "A producer marketing board may be described as a horizontal, producer-oriented organization established under government legislation which gives the board various legal powers of compulsion over producers and, in some instances, over manufacturers and handlers of primary or processed agricultural commodities, and which operate in the interest of agricultural producers."[53]

Producer marketing boards are prevalent in Australia, Canada, New Zealand, and the United Kingdom. The United States has never adopted the use of producer marketing boards. Many researchers question this behavior since Canada, which borders the United States and has quite similar agriculture, makes extensive use of marketing boards. Canada uses boards widely "probably in excess of 80 percent of the value of production by primary producers."[54]

Some researchers do foresee possible uses of producer marketing boards in the United States for certain agricultural products. Boards would have the impact of improving information available to producers and the public and improving the balance of power between producers and buyers.[55]

Another aspect of marketing boards is that a number of industries or consumer groups can be represented, depending on who is elected or appointed to oversee the functions of the board.

Most core sample managers were undecided whether marketing boards have any potential as a major part of United States agricultural cooperatives' strategy in the next decade. The responses from the academic personnel were evenly divided on this same issue.

TAXATION. Cooperatives like other business corporations are subject to many kinds of taxes including federal corporate income taxes. But cooperatives, depending on how they operate, are treated differently in the federal tax area.[56] It is in the federal tax area that public controversy has raged over so-called cooperative tax "exemptions."

Importance to Agricultural Cooperatives. The two important provisions of the Internal Revenue Code currently affecting agricultural cooperatives are Subchapter T and Section 521. Cooperatives qualifying for either of these tax provisions must follow certain prescribed rules that restrict how they acquire and distribute net margins.

Subchapter T is the essential feature of the cooperative tax treatment and applies to "*any* corporation operating on a cooperative basis." *No* requirement holds that the corporation must deal in agricultural products. Basically, the law says cooperatives are not subject to tax at the corporate level on net margins derived from business with patrons if these margins are distributed or allocated to patrons on the basis of patronage and if they follow a set of comprehensive rules.[57] The concept is one of a single tax on earnings—payable currently either at the patron or the cooperative level. The belief is that the cooperative itself does not realize a profit from its operations but is simply an extension of the member's farm business. In this sense, the tax status is similar to that of a partnership.

Subchapter T of the Internal Revenue Act of 1962 states that to be taxed only at the patron level a patronage dividend must:

1. Be computed on the basis of quantity or value of business done with patrons

2. Be made pursuant to a contractual obligation that was in effect before any transactions took place with patrons

3. Be determined by reference to net earnings of the organization from business with or for its patrons

4. Be paid in cash, property, or qualified written notice of allocation

5. Be paid within eight months following the end of the cooperative's fiscal year

For the patronage dividend to qualify as tax deductible to the cooperative, at least 20 percent of the amount must be paid in cash.[58] Furthermore, patrons must consent to report the noncash portion of their patronage refund as current income. This has caused some widespread problems, especially in years of high farm income. Farmers in higher income tax brackets often find that the minimum 20 percent cash payout is not sufficient to cover their tax liability.

Section 521, available only to agricultural cooperatives, establishes the requirements that cooperatives must adhere to in order to qualify for certain deductions set forth in Subchapter T.[59] Subchapter T allows Section 521 cooperatives to deduct not only patronage refunds but amounts paid as dividends on capital stock and amounts of income received from business with nonpatronage sources such as the United States government, rentals, and interest.

Qualification is based on a set of rigorous requirements, the most important of which are:

1. The organization must be a true farmer organization organized and

operated on a cooperative basis. "Substantially all" capital stock must be owned by producers. Records and net margins for marketing and supply activities must be kept separately.

2. At least 50 percent of dollar volume for marketing and supply cooperatives must come from members. In a supply cooperative, purchasing for persons who are neither members nor producers must be limited to 15 percent of the total.

3. The dividend rate on stock must not exceed the greater of 8 percent or the legal interest rate in the state of incorporation.

4. Financial reserves are restricted to those required by state law or those reasonable and necessary. In almost all circumstances, they must be allocated to patrons.

5. Nonmembers must be treated the same as members in pricing, pooling, or payment of the sales proceeds, and in the allocation of patronage refunds.

Tax-Related Issues. Certain of the preceding requirements have been further defined in the course of numerous Internal Revenue Service (IRS) rulings in such a way as to narrow the restrictions for cooperatives operating within the requirements of either Subchapter T or Section 521. These rulings may affect cooperative growth.

OPERATING ON A COOPERATIVE BASIS. The IRS has attempted to make the definition of "operating on a cooperative basis" more explicit.[60] This phrase is contained in both Subchapter T and Section 521. The IRS has required that a cooperative must do at least 50 percent of its business with members to be considered as operating on a cooperative basis. The 50 percent rule is spelled out clearly for cooperatives satisfying Section 521 requirements. However, for those cooperatives not satisfying Section 521 requirements and whose nonmember business is approaching 50 percent of the total, the 50 percent interpretation may be an important new restriction.[61] If member business dropped below half, none of the patronage refunds would be excludable. For those cooperatives selling farm supplies in or near urban areas, this IRS restriction may be of particular importance.

Another important restriction bearing on the "cooperative basis" interpretation is the one-member, one-vote issue. In a technical advice memorandum of May 30, 1974, the IRS suggested that a Section 521 status be revoked because the cooperative provided for voting rights other than on a one-member, one-vote principle.[62] Action on this issue should be closely watched by those cooperatives having voting rights based on volume or past patronage.

A strict view of the "operating on a cooperative basis" principle has led the IRS to challenge cooperative practices in two other areas. One is that of cooperative losses, the other is intracooperative netting of margins and losses. Recent tax court cases have tended to support the cooperative viewpoint on these issues.

The IRS argument on losses is that the principle of nonprofit operation prevents a "true" cooperative from having a loss for tax purposes. When a cooperative has a net margin it is obligated to return that margin to patrons on a patronage basis. If the expenses and advances exceed proceeds, a "symmetry" of argument would require that those losses or "negative net margins" also be allocated to members and patrons. The cooperative would either have to reduce recent equity accounts, reduce future patronage refunds, or, ideally, collect more from patrons to bring the net margins up to zero. In two recent cases the tax court has rejected this line of reasoning. However, there may be continued pressure by the IRS to enforce this concept of cooperative operation.

The strict view of cooperative operation has also led the IRS to object to some netting practices within cooperatives. When one department suffers a loss and another department of the same cooperative has a net margin, the practice is generally that the members, acting through their boards of directors, may, if they so desire, spread the loss and margins among all members. The objection to this is that "equity" principles are not properly observed. The cooperative position is generally that the cooperative itself is the tax-paying unit and the IRS should have no interest in the internal operations of the organization. A recent tax court case has specifically permitted netting among departments.

If the strict IRS view prevails on these two issues, a major reevaluation of operating methods would be required. All cooperatives, not just Section 521 cooperatives, would be affected, and the use of cooperatives by farmers could be significantly altered. If the challenge is severe, the dispute could be elevated from an administrative enforcement issue to a legislative one.

THE LOOK-THROUGH PRINCIPLE. The look-through principle, an interpretation in effect for all tax years beginning after December 17, 1973, says that federated cooperatives must disregard the corporate status of their member locals and look through to the ultimate patron to determine whether all strict conditions of Section 521 are being met.[63] "The federated cooperative is considered to be dealing directly with the patrons of the members."[64] The principle does not rule out the possibility of a federated cooperative having nonexempt member locals, but these nonexempt members cannot engage in activities prohibited directly to the exempt federated cooperative.

CURRENT BUSINESS. Part of the language of the Section 521 provision has been interpreted as a requirement for producer members to be active or current. "The revenue service has taken the position that a producer will not be considered an active producer unless the producer either currently markets more than 50 percent of the particular product through the cooperative or purchases more than 50 percent of his supplies from it during the year."[65] Local and centralized cooperatives find this requirement impossible to meet—not necessarily because of ineligibility, but because they have no way of knowing whether a member purchases 50 percent of his supplies or markets 50 percent of his product through the cooperative. Further developments in this area are likely as the IRS becomes more aware of the difficulties of enforcing the requirement.

SECTION 521 STATUS. Although up-to-date numbers are not available for all cooperatives, two previous studies give an idea as to the number of cooperatives that do not have Section 521 status. Nelda Griffin, in a study using 1976 data, found that 57 percent of 5,795 cooperatives surveyed did not have Section 521 status.[66] Charles Kraenzle and Francis Yager found in a study of 1,367 local grain cooperatives that 644 or 47 percent did not have Section 521 status.[67]

Cooperatives in this study were asked about their Section 521 tax status: whether they had the "letter of exemption"; if so, why they wished to retain it; and if not, why they had given it up.

Most of the large regional cooperatives studied had given up their Section 521 status—some quite recently. One reason appeared to be the difficulty in meeting the member or producer patronage requirements. Cooperatives selling farm supplies stated that they needed to sell to nonmembers and nonproducers to gain the volume required to serve farmers in the area efficiently. Marketing cooperatives often relied on nonmember producers to round out raw material supplies needed for efficient processing plant usage. Under Section 521 these nonmember producers would have to be treated the same as members. Several federated cooperatives had given up their Section 521 status because their locals had given up theirs (a result of the look-through principle).

Cooperatives having given up the Section 521 status pointed to several advantages the new status gave them. Several managers cited the high legal and accounting fees required for keeping the exemption. In some cases this was why the status was originally relinquished. A few believed the lack of Section 521 status improved their image with noncooperative firms and with commercial lenders. Apparently the "exempt" wording was sometimes responsible for creating an unfavorable image for cooperatives and for setting them apart in financial and business circles. Probably the most common advantage expressed by managers of those having given up Section 521 was the additional flexibility it gave them in running their business. Fewer restrictions were now placed on the way they operated.

Those cooperatives that qualified for Section 521 believed that tax savings justified the restrictions IRS regulations forced upon them. One manager said, "We fight every day to keep it [the status] but we find it is still more advantageous to keep it than give it up." In this case the dividend exclusion provided considerable savings that could be allocated to patrons. One cooperative justified its Section 521 status by saying, "Section 521 is a safeguard to make sure there is no distinct separation between the goals of management and the goals of growers. It keeps the cooperative operating according to what the charter says it has been organized to do." The safeguard appeared to be the Section 521 restrictions imposed on nonmember, nonproducer business.

Although not mentioned by cooperatives surveyed, Section 521 status for many cooperatives may be more important because of the exemption allowed from Securities and Exchange Commission (SEC) registration requirements than from any tax savings.

Section 521 status is an important decision that bears directly on cooperative growth patterns. The decision goes beyond simple calculation of tax benefits. Section 521 regulations restrict cooperative growth in the area of nonmember, nonproducer business; they determine treatment of member versus nonmember net margins; and they even have an effect on overall management philosophy. Section 521 status also enhances cooperative growth through savings in federal taxes.

No attempt will be made here to recommend either the "exempt" or the "nonexempt" strategy. The decision must be made on a case-by-case basis. It is possible, however, to identify certain key factors that affect cooperative strategy in this area. Cooperatives with the following characteristics may want to maintain the Section 521 exemption:

1. Cooperative wishes to obtain new members and increased volume in a systematic manner through use of patronage refunds.

2. Operation does not require doing business with many nonmembers who are nonproducers.

3. Significant amount of dividends paid on capital stock.[68]

4. Capital requirements depend on public stock or debt sales that might, for some cooperatives without the status, require the burden of compliance with SEC regulations.[69]

Cooperatives with the following characteristics may not wish to obtain or retain a Section 521 status:

1. Operation usually requires doing a significant amount of business with nonmembers (or for supply cooperatives with nonproducers as well).

2. Have little or no capital stock on which dividends are paid.

3. The organization wishes to treat nonmember patrons differently than members are treated.

Section 521 requirements appear to be the cause of most cooperative problems with the IRS. Interpretation of the law has become increasingly narrow even to the point of defining what a "true" cooperative is and how it should operate. Many cooperatives studied are undergoing lengthy investigations of their past tax returns with millions of dollars hanging on the outcome of IRS decisions regarding past business activity. The "letter of exemption" is a necessary but not sufficient condition for Section 521 treatment. Cooperatives with the "letter of exemption" can have their status redefined at any time. The more stringent IRS rules have probably lessened the usefulness of the exempt status.[70]

Cooperative managers and board members who have the "letter of exemption" should keep themselves well informed on Section 521 requirements, should keep up with court cases that may affect their type of cooperative, and should review periodically their eligibility. The regulations have too great an impact on operations and ultimately on direction of growth for them to do otherwise.

Legislative Policy of Cooperatives. Many agricultural cooperatives rely on the National Council of Farmer Cooperatives (NCFC) for leadership on tax issues. The council tax policy is dedicated to supporting the single-tax principle ''as basic to the proper tax treatment of farmer cooperatives and their patrons.'' Policy is also directed at opposing legislation that is inconsistent with the single-tax principle. The NCFC ''will also oppose any proposals for limitation of statutory deductions and exclusions available to farmer cooperatives under existing law.''[71]

The managers were united in their desire to see the Subchapter T legislation maintained. However, viewpoints on the subject of Section 521 varied. Some of those having given up their status believed the legislation, with its "exempt" vs. "nonexempt" language, damaged the public image of cooperatives. The damage to their image and the increasingly difficult requirements for keeping it were not thought to be worth the amount of tax money saved. Some believed it impeded their flexibility in expanding.

Many cooperatives still have the Section 521 letter of exemption. These cooperatives because of Section 521 are able to return more money to their patrons. They are also exempt from SEC registration requirements. One manager pointed out that Section 521 may discourage undue concentration on nonmember, non-farmer business.

Cooperatives, through the NCFC, are currently presenting a united front on the subject of Section 521, but differences among cooperatives provide a question for the future.

Political Climate. There will be continual pressure to change the federal tax laws so that cooperatives cannot exclude patronage refunds from their taxable income. Cooperative tax laws, especially Section 521, are believed by many to be unfair. The desire to tighten up on tax legislation affecting nonprofit groups and to make the tax laws more uniform adds fuel to the fire. This climate could become strong enough to affect adversely future cooperative tax legislation. Most of the managers interviewed recognized this danger and were especially concerned about losing their single-tax status.

The most intense congressional pressure will come from the National Tax Equality Association and from businesses competing directly with agricultural cooperatives. Competitors often see cooperatives as having an unfair tax advantage allowing them to accumulate untaxed "retained earnings," whereas other corporations are taxed on corporate net income and their stockholders are taxed again on the dividends they receive.

Activity on the legislative front has been fairly quiet since the Tax Reform Act of 1969 was passed. That bill originally proposed an increase in minimum cash payout on allocated dividends from the 20 percent level and also proposed a maximum revolving fund period for qualified tax deductions. Both of these provisions were effectively opposed by cooperatives. Bills introduced since then have been directed at repealing Section 521 and Subchapter T provisions of the Internal Revenue Code. None of these has made much headway in Congress. A typical bill is H.R. 1790, introduced by Senator Burke of Massachusetts, which would impose a dual tax on the earnings of farmer cooperatives.

Another tax initiative has also been introduced in Congress. It proposes a "single tax" approach to corporate dividends. If passed, this legislation would certainly diffuse the arguments against deduction of cooperative patronage refunds.

Although cooperatives are not now faced with any serious legislative threats in the tax area, they are likely to keep up their vigilance primarily through the lobbying efforts of the NCFC. They will probably also put forth greater effort, both individually and as a group, toward educating the public to the legislative intent behind the cooperative tax laws and the basic principles of cooperative taxation.

In the near future, cooperative taxation will face its most severe test, not in the legislative area, but in the field of tax administration. Federal agencies have a considerable effect on cooperative taxation through their use of regulations applicable to existing laws. This is where cooperatives should look for any changes in taxation over the short run.

Using Information Systems

Availability of market information is considered a crucial requirement for a competitive marketing system. In practice, however, market participants never have full information. New techniques are evolving for collecting and disseminating information. Yet much concern exists about the level of marketing information available. Information needs have increased, price-making points generating price data have declined, firm secrecy about information has increased, and efficiency of public information systems has declined.

The country livestock auction is an example of a market where information regarding transactions is readily available to all participants willing to visit the auction market. Information available includes quantities supplied, quantities demanded, and prices paid for various qualities. However, even in this public market, information is lacking on production and processing costs, demand, supply, price conditions on private markets, and projected market conditions. The buyer on a market such as the public livestock auction generally has access to much greater market knowledge than does the local seller. The buyer can often use this knowledge to make his purchases on the auction market at the most opportune times.

Agricultural marketing practices are moving rapidly away from the open market system as practiced in the typical auction market or the historical centralized wholesale market. The evolving marketing systems are associated with vertically integrated systems, larger marketing firms, and economic concentration. The large vertically integrated firms have absorbed some of the previously open market pricing points as transfer prices between their own multi-stage operations. Market information previously public is becoming internalized within these companies. The integration is often carried back to the farm level through production contracts made privately between buyer and seller well before harvest. These contracts eliminate another large portion of the product volume previously traded on open markets. No formalized reporting system exists for reporting these contract negotiations. Market participants seeking contract price information are left to rely on USDA market news reports reported only for some commodities and only on a voluntary basis, or on market information they have encountered through their own trades.[72]

Cooperatives, as a part of this changing marketing system, are often responsible for innovations to improve the system. The farmer may find his marketing cooperative more and more essential as his own diminishing sources of market information reduce his ability to make intelligent marketing decisions for his products.

Each of the three general cooperative marketing strategies has need for good market information. Three examples of marketing information systems that relate to each general strategy will be discussed.

MARKETING INFORMATION SYSTEMS IN USE. The first information system to be discussed is Telcot. It is used by Plains Cotton Cooperative Association (PCCA), Lubbock, Texas, a cooperative mainly oriented toward maintaining a competitive market.[73] The second is used to improve the bargaining strategy and is operated by American Agricultural Marketing Association, Chicago, Illinois, an affiliate of American Farm Bureau. The third system represents the integration and coordination strategy and consists of a combination of private information systems, internally generated information, and publicly available information.

Telcot, operated by the PCCA, is an electronic marketing system available for farmers' direct use in obtaining marketing information and in offering their cotton for sale. Terminals are located at the farmers' local gin offices. Farmers can use the terminal as a market information service to give them current sales of varying cotton qualities throughout their trading areas. If they think the market is right they can offer their cotton for sale over the terminal, setting their own asking price, and alternatively soliciting bids from cotton merchants (including the PCCA) tied into the system. Buyers compete for the cotton by making bids on terminals located in their own offices.

The Telcot system contributes to the maintenance of a competitive market by providing two important services. First, the availability of market trading information is increased for all segments. Farmers can readily discover actual trading prices for cotton with characteristics similar to their own. Second, Telcot provides an efficient electronic exchange system that brings both local and distant buyers together to compete for cotton offered for sale, thus providing the farmer a ready market. The participation of the PCCA in the bidding also helps enhance the competitiveness of the market.

The American Agricultural Marketing Association serves as the national coordinating and information agency for about forty-five state Farm Bureau Marketing Associations. In this capacity the AAMA acts as a marketing agency in common for its member associations. Participating state associations market a variety of products with special emphasis given to processing fruits, vegetables, broilers and other fowl, and livestock. The AAMA reports specific contract terms as sent in by each of its state affiliates. The association in many instances recommends minimum prices farmers should accept when signing production contracts. In this way the AAMA provides a nationwide information service that arms each member bargaining association with a better knowledge of contract terms as negotiated by member bargaining associations in other states.

In the bargaining process, information availability is crucial to the design of an effective bargaining strategy. The information system described above may be quite helpful for providing information on contract terms as well as some information on general supply and demand, but the individual bargaining association must also be well informed on a variety of issues specific to its locality and market. The bargaining association should conduct its own economic analysis concerning information such as trends in crop size, yields, quality, product demand, cost of production, and even cost-of-living indexes. One bargaining association manager said he used fourteen different factors in arriving at a formula for determining contract prices.

A bargaining expert interviewed indicated that "the biggest problem facing bargaining cooperatives is getting information. The processors won't give them any information because they think bargaining cooperatives are their business enemies and producers often do not realize the importance of gathering information." Another expert voiced a similar opinion: "You must have resources to get the relevant information; this is part of the bargaining association's responsibility. Farmers must be willing to pay the price for the necessary information."

The cooperative using the integration and coordination strategy should have its own market analysis staff. These people would have access to internally generated information resulting from the firm's multi-stage operation and to information derived from sales of its processed products. Additional information could be supplied from public market news reports and from subscriptions to private information-gathering services. The cooperative could also participate in an industry-wide trade association that collected aggregate marketing information from its membership. If a federal or state marketing order were in existence for the commodity, the cooperative could also utilize marketing data collected by the order. The integrated cooperative might even participate in a market information cooperative designed to provide a forum for information exchange. The cooperative's economic analysis staff would have the responsibility of putting all this information together, analyzing it, and providing interpretations helpful to the firm's decision makers.

ACQUIRING AND USING MARKET INFORMATION. Core sample questionnaires indicated almost unanimous agreement that large regional cooperatives had no substantial problem in acquiring useful market information for farm product marketing. Another question asked cooperative managers to compare their market information with that of their suppliers, their competitors, and the buyers of their products. Almost all agreed their market information compared favorably with that of their suppliers (whom in most instances would be their grower members), but only a slight majority felt that way about their competitors' or buyers' information. The remainder were undecided on the issue.

The cooperative managers were then asked whether the ability of their staff to use market information compared favorably with their suppliers, competitors, or buyers. All agreed their use of market information compared favorably with that of their suppliers, almost all agreed their use was favorable with that of their competitors, and only a simple majority agreed with respect to their buyers. Academic personnel were asked the same question with respect to agricultural cooperatives as a group. A slight majority agreed that the ability of the staffs of cooperatives to use market information compared favorably with that of suppliers and competitors, but there was no consensus with respect to buyers.

USEFULNESS OF ALTERNATIVE INFORMATION SOURCES. At least a majority of managers agreed that the following were excellent information sources: internal information on the general environment, USDA series, trade associations, and trade magazines. A few cooperatives also believed private subscription series such as the American Institute of Food Distribution and Kiplinger were excellent sources.

Cooperatives were undecided about multiple cooperative organizations as an information source. The majority disagreed with the idea that joint ventures with noncooperative firms provided an excellent source of market information.

Both cooperatives and academic personnel were asked what constitutes excellent programs for providing market information to agricultural cooperatives as a group. Cooperative respondents reached no consensus with regard to the usefulness of multi-cooperative information systems, subscriptions to private systems, or joint arrangements with noncooperative interests. However, almost all cooperative respondents agreed that excellent programs would include encouraging improvements in available government data and developing a better information base within their own organizations.

A majority of the university personnel believed that excellent programs would include establishment of a multi-cooperative information system and subscriptions to private systems. University respondents were evenly divided on the use of joint arrangements with noncooperative firms. Like the cooperative respondents, almost all agreed that improvements in available government data should be encouraged and that a better information base within individual organizations should be developed.

ASSESSMENT. Cooperatives studied did not believe they had a substantial problem acquiring or using market information, nor did most appear concerned about their abilities in this area when compared with their suppliers, competitors, or buyers. These results, even with regard to the large regional cooperatives, were surprising. For those cooperatives operating in markets dominated by much larger noncooperative firms, the attitude may even be somewhat naive. These larger companies usually have access to more information and to larger, more highly trained staffs of economic analysts prepared to interpret market data for the firm's private use. These organizations may also have closer knowledge of consumer demands. Their contacts and influence with legislative bodies—both state and federal—may be considerably greater than those of the typical cooperative.

Cooperative managers surveyed did not appear to be thinking much about multicooperative efforts in the information area. Instead, their attitude was similar to what might be expected of a noncooperative firm—development of internal systems, reliance on government data, and use of the industry trade associations. These information sources are certainly important, but cooperatives might also consider whether their cooperative status provides them with any special opportunities in the information area. One example is their legal right to form a marketing agency in common for information exchange. This was done by the AAMA.

Cooperatives also indicated an interest in improving available government data. One approach to this involves mandatory public reporting of market transactions.[74] Publicly available government information would then be more credible and much more complete than information based on voluntary disclosure. Cooperatives as a group may have a responsibility to study approaches such as these. If found of value, cooperatives should push for implementation.

Enhancing Domestic Demand

Domestic demand enhancement is included as a facilitating strategy because of the supporting role it can play for a marketing cooperative, especially for the integration and coordination strategy. Activity in this area may be particularly important, given the nature of today's farm programs and today's dependence on volatile foreign markets. For farmers, these factors have contributed to the volatility of their business. Farmers need the support of their cooperative which, although it has no control over supply, can attempt to influence the demand for farm products and improve farm income.

An increasingly important part of demand enhancement may fall on cooperatives because "old line" food companies have diversified and conglomerated, finding profitable alternatives to the food business. They are no longer as interested in stimulating demand for agricultural products. Furthermore, farmers can rely only to a limited extent on their participation in industry-wide trade associations or producer checkoff programs to enhance demand. Their cooperative, however, may be in a position to transform raw farm products into products that can be differentiated in the consumer market, thus creating a higher potential demand and a higher market value.

PRODUCT PROMOTION AND DEVELOPMENT. USDA researchers found that among 1,172 agricultural groups in fiscal year 1962-63, a total of $86 million was spent on farm product promotional activities.[75,76] Cooperatives spent $27 million or 31 percent of this total. Other groups spending funds for this purpose included voluntary producer-processor groups spending $31 million or 36 percent and commissions, boards, councils, and institutions which in the aggregate spent $25 million or 29 percent. The remaining 4 percent was spent by state departments of agriculture and other unidentified groups. Preliminary results from the 1973 survey (1972 fiscal year) indicate overall expenditure of $139 million by 917 groups, with cooperatives spending 28 percent. Taking out the effects of inflation since 1963, the $135 million (average) figure amounts to only an 11 percent increase in expenditures for all agricultural groups.[77]

Managers of cooperatives were questioned as to whether they expected adequate effort to be given to farm product promotion or product development over the next decade. The majority were undecided on both issues. A few did agree that the effort would be adequate. Most of the university experts, however, agreed there would be adequate effort for product promotion; a majority believed the same about product development.

The managers were also asked about the responsibilities of cooperatives in the area of product development and product promotion during the next decade. All managers agreed that cooperatives should increase their responsibility in these areas. Apparently, through cooperative action, they hoped to correct some of the uncertainty they expressed in the first two questions. Academic experts, however, reached no consensus on the issues.

Promotional activities by cooperatives are carried out almost solely by cooperatives organized for the integration and coordination strategy. Cooperatives having integrated forward to the consumer level with their own differentiated brands stand to benefit most from the expanded demand they can create through their own promotional efforts. Preliminary estimates from the fiscal 1972 USDA survey indicate that 46 percent of all cooperative expenditures were for promotion of branded products, while 35 percent were spent for unbranded products that were identified with an area of origin.[78]

Only a few cooperatives account for the majority of cooperative expenditures for advertising of farm products. These include the following:[79]

Name	Headquarters
Ocean Spray Cranberries, Inc.	Plymouth, Massachusetts
Sunkist Growers, Inc.	Van Nuys, California
Cal. & Haw. Sugar Assn.	San Francisco, California
California Almond Growers Exchange	Sacramento, California
Diamond Walnut Growers	Stockton, California
California Canners and Growers	San Francisco, California
Sun-Maid Raisin Growers of California	Kingsburg, California
Sunsweet Growers, Inc.	Stockton, California

These cooperatives market specialty crops, and they have gained over time a market premium for their high and uniform quality products.

Cooperatives such as Tri-Valley Growers and Citrus Central, Inc., have concentrated on marketing private label products with a small percentage sold on a regional basis under their own brand name. Cooperative use of the private label is quite common. Lack of advertising funds is one reason for the use of private labels. Another reason is the desire to avoid the financial risk associated with brand advertising that competes directly with the advertising of national food companies.

An assortment of products produced for sale under cooperative brand names.

Cooperatives such as Pro-Fac (through Curtice-Burns) and Riceland Foods have concentrated on promoting their consumer products through regional brands. Regional branding is of special interest to cooperatives because their production base is usually regional and because of the limited funds usually available for brand advertising. Cooperatives concentrating advertising dollars on regional brands where they are more familiar with the market can have, with limited funds, a greater effect on enhancing demand for their product.

Although some cooperatives have been successful in promoting their own brands, most managers did not believe that cooperatives as a group were well prepared for promoting either regional or national brands. University experts agreed, and they believed that cooperatives should put greater emphasis on marketing their own branded products. But since most cooperatives are not in a position to market either branded or nonbranded products directly to the consumer they have little incentive to engage in promotional activities.

Marketing cooperatives with a well-established label have generally not seemed willing to share or franchise their label with other cooperatives in multi-cooperative arrangements. However, two examples of multi-cooperative sales agencies set up to market the products of member cooperatives under the members' own separate brand names are Citrus Central, Inc., Orlando, Florida; and California Valley Export, San Francisco, California.

Another common approach is for the multi-cooperative sales organization to market brand names not previously identified with any organization. Sunkist and Citrus Central have done this for their federated packing houses. This approach may be worth more consideration by cooperatives lacking the financial resources to promote their own brand.

Although advertising of branded products represents the greatest part of cooperative promotional expenditures, advertising of nonbranded products represents the greatest part of expenditures made by agricultural groups as a whole. The USDA survey found that an estimated 81 percent of the 1972 fiscal aggregate promotional expenditures were for nonbranded products.[80]

Agricultural producers and their cooperatives rely on several methods for expanding the demand for nonbranded (or generic) farm commodities. One method is the mandatory checkoff at the producer or processor level. In the cotton industry, the $1 per bale checkoff finances an $8–$12 million annual program funding research, product development, and promotional activities. Cotton Incorporated, a producer-controlled group, operates the program.[81] A top officer of Calcot, a cotton marketing cooperative in California, stated that for product promotion and development "we look to Cotton Incorporated which has the overall responsibility to influence the use of cotton through to the retail level."

Many of these generic research, developmental, and promotional programs operate under the authority of state or federal marketing orders. This is the case with milk and many of the fruits, nuts, and vegetables. The California Tree Fruit Agreement is an example.

Producers also share promotional expenses through voluntary trade associations such as United Dairy Industry Association (UDIA) or the American Soybean Association. Sometimes processors and handlers share expenses as members of these associations. State departments of agriculture also play a role in promoting farm commodities, especially products that can be identified as being from a specific state.

A potential area in which the demand for farm products can be enhanced is the development and promotion of *new* food products that are derived from raw farm products. Cooperative new product development and promotion activities were not probed in detail, but some results can be reported. Cooperative managers and academic experts were asked if they thought certain organizations were well situated for developing new products from raw farm products. Both groups were unanimous in agreeing that noncooperative business firms and private brand firms were well situated for doing this.

A majority of the cooperative respondents believed land-grant universities were well situated to develop new products, but university experts were almost evenly divided on that issue. The USDA was seen as being well situated for developing new products by the majority of the academic experts, but cooperative managers reached no consensus on this point.

The two survey groups were asked the same question with regard to the *promotion* of *new* products. Almost all agreed noncooperative and private brand firms were well situated for promoting new products. The majority of cooperative managers disagreed, and nearly all college experts disagreed, that land-grant universities were well situated. Both groups agreed that the USDA was not well situated, but a majority of both groups agreed that cooperatives were well situated for promoting new products.

The majority of the managers surveyed were undecided about the prospects for the success of voluntary producer checkoff programs for new product development or promotion. However, in a separate question the majority stated they did participate in producer checkoff programs. Almost all the university experts surveyed agreed that these voluntary programs have only limited chances of success for new product development.

The sample cooperatives were asked about their participation with other producer groups or noncooperative concerns for the purpose of product development or promotion. The majority indicated they did participate in such groups. These groups included the American Egg Board, Cotton Incorporated, National Broiler Council, National Turkey Federation, American Soybean Association, and the Rice Council for Market Development.

Most managers agreed with the statement that a bright future exists for new uses of farm-produced raw materials. Academic experts were divided on the issue although a majority did agree with the statement.

SYNTHETICS AND SUBSTITUTES. Synthetics and substitutes for agricultural products have already caused some significant shifts in consumption patterns. Cotton has faced the most dramatic competition from nonfarm synthetics. Currently, cotton fiber and wool together represent only about 30 percent of the fiber used in the domestic textile industry.[82]

Margarine has captured two-thirds of the table-spread market. Synthetics or substitutes have captured over 21 percent of retail citrus beverage purchases. Other foods and beverages including whipping cream, sweeteners, and dairy creamers have lost substantial sales to new products.[83]

Currently the substitution of vegetable protein for animal protein is receiving considerable attention. Advancements in technology and processing are making vegetable proteins more like meat and dairy products in texture, flavor, and nutritional characteristics.[84] In the red meat area the major market penetration has been in processed items through the use of less costly soy protein extenders.[85] A 1972 study projected that between 10 and 21 percent of the processed meat market would be replaced by soy protein by the year 1980.[86] This level of substitution will be a result of not only the projected lower cost of the soy protein but also the adverse publicity over the use of animal fat in the diet. The potential for using this inexpensive high protein source to satisfy nutritional deficiencies among the world's underfed people will be another significant factor in its use as a substitute.

Survey participants were asked how much substitution they expected in the use of plant for animal protein during the next five and ten years. About one-half of the cooperative respondents predicted a slight shift and half predicted a moderate shift over the next five years. Several respondents and all the academic experts believed that the shift would become greater over the ten-year period.

Cooperatives that have developed research and sales efforts may need to adjust to shifts from animal to plant protein. Cooperatives could easily expand their research efforts in the soy protein area and develop their own patented technology. In this manner, cooperatives might be able to capture a share of the expanding markets.

Cooperatives also must consider the threat of food and fiber substitutes from nonfarm sources. Petroleum or coal derivatives are the most likely of these synthetics. The inroads made by the synthetic fiber industry into the markets for cotton fiber serve as a dramatic example. Food derived from the sea may also be a factor in the future.

The majority of managers did not think that nonfarm food and fiber substitutes would become a stiffer competitor to farm-based food and fiber during the next five years. But they were evenly divided when asked about the prospects over the next ten years. The opinions among university professors were arrayed in a similar way for the five-year projection, but the majority agreed that nonfarm substitutes would become stiffer competitors over the next ten years.

For the long run, agricultural groups must ask themselves a crucial question: Do farmers stand to lose a significant portion of their markets to nonfarm food sources if they do not more actively promote their existing products and at the same time develop new products to fit changing consumer needs?

ASSESSMENT. The individual producer, because he produces a nondifferentiated product, must seek the company of other producers if he is to justify product promotion expenditures. If this is done through a trade association or state agency, there will be no control exercised over product marketings or quality except in the case of some commodities under marketing order legislation. A group such as this cannot, without ownership of the product, exert any control over product pricing, packaging, or other factors affecting product sales. Without any merchandising control, an increase in product demand will not necessarily guarantee increased producer returns. Furthermore, the trade association or state agency will find it extremely difficult to measure the direct effects of its promotional activities even if it can measure a sales increase for the product. One of the most severe problems will result if the group is actually able to improve the demand and to raise prices. The supply response from producers, whether they are members of the group or not, may adversely affect prices in subsequent periods. Despite these problems, commodity advertising through producer groups has been a major marketing tool for milk, poultry, and citrus producers.

Demand enhancement, if carried out by a cooperative, can be treated as part of an integrated total marketing plan. Although the cooperative seldom has the control over supply that a noncooperative firm has, it can, through product promotion of branded products, coordinate such marketing variables as quantity and quality marketed, product pricing, market timing, and packaging with its sales effort to maximize returns. If the product can be sufficiently differentiated, results of promotional efforts can be captured for cooperative members, although the cooperative will be susceptible to its lack of control over raw material supply.

If the product is generic in nature, it is probably difficult even for the cooperative to justify promotional efforts. The job might better be done through a checkoff program involving overall industry cooperation if the effort can be adequately funded and effectively administered.

Questionnaire results made evident the opinion that noncooperative firms were well prepared for product promotion and product development. Cooperative potential, at least in the area of product promotion of branded products, was not considered good. Agricultural producers and their cooperative organizations need to ask themselves whether they can afford to allow responsibility for effective promotional activities to remain in the hands of firms with no direct ties to agricultural producers.

Expanding Foreign Trade

Expected foreign demand is a key factor in choice of general strategies for cooperatives. Just as many aspects of domestic demand enhancement are important, so also are elements of foreign demand. Many believe that more involvement of cooperatives relative to the large noncooperative firms in foreign trade would be economically and politically advisable. As the United States government food export programs evolve, the place of United States cooperatives in such programs becomes an increasingly relevant question. It is for these reasons that foreign trade expansion issues are considered as a facilitating strategy.

Farm exports have become of great importance to American agriculture and to the American economy. In the last six years value of total domestic agricultural exports has tripled—from $7.3 billion in 1970 to $23.0 billion in 1976.[87]

Several important questions arise with regard to the future of foreign trade. Should agricultural cooperatives directly export more farm products? Should they become more influential in foreign trade policy and if so, how should it be done? Are cooperative leaders united enough to influence United States foreign trade policy? Can cooperative organizations be more effective in foreign trade through joint efforts rather than by independent efforts? Do noncooperative organizations and others who are not primarily concerned with farmers' interests have greater influence in foreign trade?

SHIFT IN IMPORTANCE. Since 1972 United States exports of agricultural products have been an important part of United States agricultural policy. Current agricultural policy is dependent on continuing demand from foreign countries for larger quantities of United States agricultural products. It is only through continuous sales of certain products in overseas markets that farmers can continue to obtain a favorable price for the products they produce without government subsidization or without reducing the quantity produced.

The greater involvement by American farmers in the world market, however, is not without new problems. There is greater uncertainty with regard to future demand. Prices of agricultural products have become more volatile. American farmers can no longer concern themselves only with those variables influencing the domestic market. Now they are affected with what is happening worldwide. Cooperatives' strategies must reflect this. These strategies also must take into account the operational differences in exporting as compared to domestic marketing.

All cooperative managers and most academic personnel believed that the sale of United States agricultural products to foreign countries would continue to increase over the next five to ten years, but the managers were much more optimistic about future foreign sales than the academic personnel.

In the future, foreign trade in processed farm products may be more important. A majority of the managers believed this to be true. One-third, however, were undecided. The response by academic personnel on this particular subject was not as positive as the response received from managers.

According to the majority of core sample managers, United States agricultural cooperatives are inadequately organized to be a major factor in foreign trade policy. Most of the academic people agreed. The problem is that cooperatives representing farmers are not involved enough in trade policy decision making, partly because farmers cannot agree among themselves on what trade policy should be. Part of the problem is that farm groups have conflicting economic interests. Livestock producers want cheaper grain while grain producers want higher-priced grain.

Is it government policy or is it the marketing aggressiveness of individual companies that has greatest influence on increasing United States agricultural exports? No consensus was shown from the responses of the managers concerning the above question. However, most of the academic people believed that the importance of United States agriculture in foreign trade depends more on government policy. Somehow, farmers through their cooperatives, commodity organizations, and general farm organizations must get together if they are going to be influential in foreign trade policy.

CURRENT POSITION OF COOPERATIVES. In a study involving cooperatives' foreign trade in fiscal year 1970, Henry W. Bradford and Richard S. Berberich found that cooperatives exported 21 percent of total United States exports.[88] In a study of agricultural exports by seventy-three cooperatives, Donald E. Hirsch found that in 1976 cooperatives accounted for 9.2 percent of total United States agricultural exports.[89]

In the study the author separated the commodities into nine different groups. During 1976 the seventy-three cooperatives' largest share in terms of value was in nuts and preparations; the fruits and preparations group was second (Table 6.1).

TABLE 6.1.   Value of cooperatives' United States direct exports, by commodity groups, 1976

Commodity group	U.S. total	Cooperatives Value	Cooperatives Percent of U.S. total
		$1,000	
Animal and animal products	$ 2,379,563	$ 34,175	1.4
Grain and preparations	10,875,277	931,549	8.6
Fruit and preparations	770,079	292,704	38.0
Nuts and preparations*	198,249	79,479	40.1
Vegetables and preparations	674,060	18,360	2.7
Feeds and fodders	448,752	10,093	2.3
Oilseeds, oilnuts, and products	5,070,368	427,157	8.4
Cotton, raw, excluding linters	1,048,669	231,664	22.1
All other	591,081	5,464	0.9
Total	$22,056,098	$2,030,645	9.2

Source: Donald E. Hirsch, "Cooperatives Directly Export $2 Billion in Farm Products," *Farmer Cooperatives,* ESCS, USDA, Washington, D.C., May 1978, p. 9.
* Excluding peanuts and products.

Many farmers are aware of cooperatives' poor position in exporting grains and soybeans. In fiscal year 1974 cooperatives exported only 25 percent of the total grain exports with only 7.5 percent exported directly. In fiscal 1976, 8.7 percent was exported directly. Although cooperatives have been increasing the volume of grain they export, the share of total United States grain and soybeans exported directly by cooperatives is small in comparison to the volume exported by the group of noncooperatives handling grains and soybeans.[90]

Cooperatives have been making greater use of multi-cooperative organizations in expanding foreign trade. Two examples of joint marketing efforts are Farmers Export Company, Overland Park, Kansas, and Amcot, Bakersfield, California.

Farmers Export Company is a federation of six regional grain cooperatives—Missouri Farmers Association, Inc., Columbia, Missouri; Farmers Union Grain Terminal Association (GTA), St. Paul, Minnesota; Farmers Grain Dealers Association of Iowa, Des Moines; Illinois Grain Corporation, Bloomington, Illinois; FAR-MAR-CO, Inc., Hutchinson, Kansas; and MFC Services (AAL), Jackson, Mississippi. The purpose of the organization is to expand grain sales overseas for member cooperatives by coordinating their export sales and shipments.

Joint efforts by the six regional cooperatives allow:
1. Ownership of facilities necessary for exporting grain
2. Greater use of management expertise
3. The volume necessary for expanded overseas sales
4. Increased earnings for member cooperatives and their farmers

Amcot is a worldwide sales agency for four cotton cooperatives—Calcot, Plains Cotton Cooperative Association, Staple Cotton Cooperative Association, and Southwestern Irrigated Cotton Growers Association. Amcot handles some domestic sales and all foreign sales for its four members through seven sales offices. Four overseas offices are located in Brussels, Belgium; Osaka, Japan; Seoul, Korea; and Hong Kong, China.

Rice being loaded for export at Port of

Advantages of such a worldwide sales agency as Amcot are:

1. Increased volume of business resulting in improved grower returns through market expansion
2. Improved market coordination
3. Elimination of duplicate sales offices and commission agents
4. Improved customer visibility and confidence
5. Increased number of cotton varieties for sales through one agency
6. Increased and better market information[91]

Sacramento Bulk Rice Facility by the Rice Growers Association of California.

Cooperatives planning joint efforts in expanding foreign trade face problems similar to those in other multi-cooperative organizations. Such problems as equitable cost sharing by members, reluctance to delegate necessary decision-making authority to another firm, lack of cooperative commitment and sometimes lack of confidence in the operations of the export agency, and delays in reaching a consensus on important decisions may hinder cooperative efforts in this area.

A few managers expressed a desire to do more in foreign trade on their own. The reasons given were (1) that their organization was large enough, (2) that they could make quicker decisions than could be made by joint efforts, (3) that going on their own would allow them to set their own pace in the direction they would like to go, and (4) that an individual cooperative had an advantage over a noncooperative organization since some foreign buyers were more willing to talk to an agricultural cooperative. They pointed out that the image of cooperatives often was better in terms of quality of product and dependability of service than that of other suppliers. "The word 'cooperative' does open doors," stated one manager.

ASSESSMENT. Cooperatives need to do more in some cases to improve foreign trade of agricultural products. Marketing of United States farm products in foreign markets is necessary to maintain adequate domestic prices. This is especially true for grains, oilseed crops, and cotton.

Cooperatives need to work more closely together in some areas. This may be the only way they can develop necessary brands, sales force, storage facilities, shipping facilities, and volume that it takes continually to compete in an international market. Often cooperative marketing firms acting alone cannot fulfill the buying requirements of the centralized buying agencies of governments in nonmarket economies.

Farmers and their cooperatives must become more vocal in foreign trade policy. Experiences of recent years have shown what effect government intervention into trade policy has on future sales and prices of agricultural products. Cooperative, commodity group, and other agricultural leaders may need to sit down together and decide what type of agricultural foreign policy would be best for United States farmers and consumers in general and make it known to those people who make trade policy decisions.

Cooperative managers who handle certain commodities can no longer consider the foreign market as a place to sell what cannot be sold domestically. Surplus dumping does not build and maintain foreign markets. One manager said, "This is the wrong way to go about it."

One cooperative expert on foreign trade indicated the importance of cooperative exports even during times of domestic shortages. His suggestion was that a minimum percentage of total United States production always be allocated to hold established foreign markets.

All but one manager and one academic person agreed that "Farmers will be at a disadvantage if their cooperatives do not become more influential in foreign trade." But, greater cooperative involvement in foreign trade of agricultural products will require greater commitment and capital investments by farmers.

Expanding Financial Sources

Cooperatives, if they wish to achieve the integration and coordination strategy or even the open market strategy, must have the foresight to plan for -capital needs.Traditional sources of capital are sometimes limited for cooperatives. Yet cooperatives also have sources of capital that are not available to noncooperative firms. Their ability to tap these sources will determine their ability to grow with their chosen strategy.

EQUITY CAPITAL SOURCES. There has been a trend toward less equity and more debt in cooperative financial structure. In 1962 Nelda Griffin showed that equity capital represented 52 percent of total liabilities for the 100 largest marketing and supply cooperatives. This compared with 29 percent for borrowed capital and 19 percent for other liabilities. Preliminary 1976 figures showed a decrease in equity capital to 34 percent, an increase in borrowed capital to 40 percent, and an increase in other liabilities to 26 percent.[92]

The Griffin report also showed that equity capital for these cooperatives in 1976 was acquired, 69 percent by retention of patronage refunds, 17 percent by per unit retains, and the other 14 percent by sale of stock or equity certificates.

Use of Revolving Capital Plans. A study of equity redemption practices by Philip Brown and David Volkin found that 71 percent of all cooperatives have some type of equity redemption program. However, only 32 percent of all cooperatives have a systematic program for redeeming equities—usually a first in, first out revolving program. Revolving fund periods averaged 10.5 years for all cooperatives that used the first in, first out revolving plan.[93]

Predominantly supply cooperatives (and some others) studied for this report had fairly long periods for their revolving fund (over ten years). Some had no revolving fund at all. Some of those without a revolving fund used other payback methods. Cash payments among these cooperatives were no more than 20–40 percent of annual net margins realized.

At the other extreme, many marketing cooperatives had board policies requiring them to pay out nearly everything in cash. Some of these did acquire their capital through per unit retains. Revolving fund cycles often were shorter than in supply cooperatives, usually less than seven years. Good reasons exist to suggest moderation of both these extremes. An American Institute of Cooperation brochure had this to say about the modern young farmer:

> [They are] well educated and profit oriented. They think big and are not afraid of taking risks to achieve their objectives. They recognize that efficiency is the name of the game and expect those with whom they do business—including their cooperatives—to be efficient. They will not patronize a cooperative out of loyalty as in the past—their attitude is more apt to be "what have you done for me lately?"[94]

W. M. Harding, former governor of the Farm Credit Administration, recently added to this idea:

> We sense a growing reluctance on the part of farmers to continue to provide equity capital and support farm supply cooperatives competing in the same territory, and unless there are distinct marketing advantages to them in belonging to large, impersonal co-ops, farmers will stay away in droves.[95]

Continued reliance on limited cash patronage refunds combined with excessively long revolving fund periods may be a policy in conflict with modern-day farmer thinking. Farmers of the future may be reluctant to invest large sums of money in their cooperatives if the cooperatives cannot provide a reasonable return on their money through higher returns for crops or readily available low-cost farm supplies and services. They may even expect that dividends be paid on equity capital.[96] A. R. Tubbs has indicated that for many aggressive farmers the capital contributions made to cooperatives are made at a large and unrecognized sacrifice to the contributing farm firm.[97] But, with the greater need for capital for facility expansion, the trend in some cooperatives may be to pay out a lesser percentage of net margins than in the past.

Some local cooperatives have had difficulty revolving the equity capital of retired farmers. This has moved some cooperatives away from the basic cooperative philosophy that those who use the cooperatives' services should finance them and that those who do not should have their equity returned. The issue has recently become politically significant and has probably hurt the image of cooperatives. Brown and Volkin found that 22 percent of the allocated equities of all cooperatives were held by inactive equity holders.[98] For many cooperatives without an equity redemption program this percentage was substantially higher. The authors found little justification for no redemption plan. The study showed that, on the average, cooperatives that did carry out formal redemption programs were in approximately the same financial position as those who did not.

Some supply regionals (Farmland Industries, for example) have developed programs to directly assist their locals in retiring equity of former members. This type of cooperation between a federated regional and its local may be an appropriate way to solve the equity retirement problem, since the regional's own patronage refund policies play such an important role in the ability of the local to retire equity.

Members do have a responsibility to adequately finance their cooperative. Farmers traditionally have not seen their cooperative as a worthy place to invest capital. The farming operation itself takes first claim on their funds. Cooperative demands for capital available for outside investment must often compete with higher returns available elsewhere. Many finance experts believe cooperatives today, especially marketing cooperatives, are undercapitalized.[99] The farmer's attitude certainly is an indication of why this is so.

The farmer should be willing to commit enough capital to guarantee the survival of a strong cooperative capable of providing needed inputs and services and/or providing an effective marketing channel. Supply cooperatives can usually generate adequate patronage capital through retention of a high percentage of their net margins. In the past, farmers have usually been willing to let this happen.

Farmers' willingness to finance their cooperatives adequately may be most critical in the marketing area. Here the equity requirements per member may be considerably greater than in the supply cooperative. "Front-end" capital, meaning capital acquired directly from the membership for a specific purpose, may be an increasingly important source of capital for new facilities. The manager of one well-known diversified marketing-supply cooperative said (when referring to the marketing area), "We can no longer rely on traditional equity funding measures. The farmer must put up the 'front-end' money." American Crystal Sugar is a good example where grower "front-end" money substantially financed a new cooperative and processing plant.[100]

One modification of the revolving capital plan used in recent years by some cooperatives is the base capital plan of financing. Capital requirements per grower are tied to recent patronage. Member equity is automatically retired if patronage declines. This financing method puts the financial burden on current patrons, maintains the necessary equity capital, and eliminates the need for the cooperative to make revolving fund payments. Several new cooperatives, including American Cotton Growers and American Crystal Sugar, have been started using the base capital plan.

As growers expect more from their cooperatives, they must also be prepared to contribute more to cooperative equity. They may be quite willing to finance those operations seen as essential, especially if benefits can be tied directly back to them. Farmers should consider investment in their cooperatives as a logical extension of their farm business. They should not think of their cooperative as a place to receive a high return on capital as they might in the stock market. If marketing cooperatives are in general too thinly capitalized, farmers must take the risk and the ultimate responsibility for their cooperatives' inability to fulfill their economic needs.

DEBT CAPITAL SOURCES. The main sources of debt capital available to agricultural cooperatives include Banks for Cooperatives, commercial banks, debt securities, and tax-free bonds. Cooperatives also use lease financing to acquire control over the use of an asset.

Griffin's preliminary 1976 data show the following shares of debt capital for the largest 100 cooperatives: Banks for Cooperatives, 57 percent; commercial banks, 10 percent; debt securities, 23 percent; other sources, 10 percent.[101] (Table 6.2.) The most significant changes in these shares have occurred in the share financed by Banks for Cooperatives and by other sources. The Banks for Cooperatives' share fell between 1970 and 1976 (from 62 to 57 percent), although it had increased between 1962 and 1970 (from 51 to 62 percent). The decline in share was offset by increases in the "other source" category. Probably most of the increase occurred in industrial revenue bonds, capital leasing, and insurance company financing.

TABLE 6.2. Sources of borrowed capital for the 100 largest farmer marketing and supply cooperatives, by major function, based on amounts outstanding at close of fiscal years 1962, 1970, and 1976

Classification and year	Cooperatives	Total borrowed capital	Banks for Cooperatives	Commercial banks	Issuance of debt securities	Other sources
	number	million dollars		percent		
Farm						
1970	16	246	63.8	18.5	10.6	7.1
1962	15	72	55.4	1.7	26.6	16.3
Marketing						
1970	62	678	68.1	10.7	16.7	4.5
1962	61	299	48.9	11.7	34.0	5.4
Marketing/farm supply						
1970	22	759	56.1	8.1	33.6	2.2
1962	21	306	50.8	6.7	35.7	6.8
Total						
1976	100	4,136	57.0	10.0	23.0	10.0
1970	100	1,683	62.0	10.7	23.4	3.9
1962	97	677	50.5	8.4	34.0	7.1

Source: Nelda Griffin, *A Financial Profile of Farmer Cooperatives in the United States,* FCS Research Report 23, FCS, USDA, Oct. 1972. Data for 1976 are preliminary and unpublished.

Banks for Cooperatives. The Banks for Cooperatives System was highly commended by the cooperative managers in this study. This access to capital was probably stressed more than any other factor as being the most important operating advantage that agricultural cooperatives have over noncooperative firms. Managers reported that the banks understand cooperatives, are vitally interested in the future survival of cooperatives as a business form, are able to offer attractive interest rates, and at times are even willing to overlook a limited equity position and offer a loan based mainly on projected earnings. In fact, some managers openly said that their existence as a successful and growing firm could be attributed to the willingness of the district bank to lend them money in times when other lenders would not. While the district bank's lending rate is often no lower than rates at noncooperative banks, it serves an important role because of the competitive pressure it puts on lending rates at other rural banks. Also, since the Banks for Cooperatives' rates are generally the same regardless of the financial condition of the borrower, smaller cooperatives or those in a weak financial position can often get lower rates than they would qualify for at noncooperative banks.

The core sample relied heavily on their district bank, especially for term debt. The exceptions to this were Farmland Industries and Agway, both issuers of large quantities of debt securities. But almost all the core sample cooperatives anticipated they would acquire a smaller percentage of their debt capital from the Banks for Cooperatives during the next five to ten years than was planned for next year.

Managers of core sample cooperatives believed the banks were progressive with regard to credit requirements and in financing new ventures.

St. Paul Farm Credit Bank Building, housing the Bank for Cooperatives, the Federal Land Bank, and the Federal Intermediate Credit Bank. This along with eleven other district Farm Credit Banks plus the Central Bank for Cooperatives makes up the Farm Credit System.

(It is recognized that many managers of small cooperatives might not necessarily have agreed with this assessment.) No consensus existed with regard to realistic loan limits or for the tendency of the banks to pursue new financing techniques.

Respondents did express concern regarding several negative aspects of the bank system.

First, a concern existed among a few of the largest cooperatives that the system tended to put such strong collateral requirements on its borrowers that it became quite difficult to seek other sources of credit. While this was of no great concern to those cooperatives wishing to borrow all funds from their district bank, it was important to those who wished to go elsewhere for a portion of their debt capital. Several cooperatives had been able to negotiate with their district bank to eliminate some of the most stringent requirements.

Second, concern was expressed about the lack of services given by the banks, especially long-range financial planning. Several managers said that one of the benefits in going to commercial banks for credit was that they were provided with a more sophisticated financial planning service, enabling better planning for long-term loans. Others said that the availability of services, such as information on foreign market potentials and export financing, was a primary reason for dealing with commercial banks.

Third, concern was expressed about system lending limits. Few cooperatives are borrowing against the system's limit now, but several of those not yet at the limit were concerned that their organization was growing faster than the system's net worth. Improved capitalization plans in many districts of the Banks for Cooperatives System should limit the number of cooperatives affected by this problem. Several managers pointed out that cooperatives, with a current or future growth rate that exceeds or will soon exceed the growth rate of the system, need to begin seeking alternative sources for a portion of their credit needs.

In 1974 a task force was formed to advise the Farm Credit Administration (FCA) on how the Banks for Cooperatives could best provide effective service.[102]

Possibly the most significant recommendation made by the task force was that "The present FCS Regulation 435(a)(1) on lending limits for Banks for Cooperatives should be retained."[103] The risk of lending more than 50 percent of the combined net worth of the Banks for Cooperatives System to any one borrower was judged to be too great.

The task force, however, recognized that the purpose of the Farm Credit System is primarily to "furnish sound, adequate, and constructive credit and closely related services to American farmers, ranchers and their cooperatives, and to selected farm-related businesses necessary for efficient farm operations."[104] The task force therefore recommended a series of steps and programs by which the banks might assist large cooperatives approaching or already at the credit limit to obtain supplemental financing.

Possibly the most important implication one could draw from the task force report was that the system did not feel an obligation to grow at a rate fast enough to accommodate the lending needs of the largest cooperatives. Yet a strong obligation was felt in assisting cooperatives to develop alternative sources of capital.

Commercial Banks. Commercial (noncooperative) banks played a minor role as a capital source in the cooperatives studied. Griffin found that only 10 percent of debt requirements of the largest 100 cooperatives were satisfied in this way in 1976 (Table 6.2). Commercial banks are used mostly for short-term or seasonal credit. Cooperatives studied, even if they were not currently borrowing from commercial banks, usually had lines of credit with one or more of them.

The primary reason for these ties to the commercial banks appeared to be the additional services they could offer as compared with those available from their district bank. Services included foreign market analysis, letters of credit, long-range planning and budgeting, a sounding board for economic projections, customer credit ratings, investment advice, and a depository for funds.

The foreign-related services were especially crucial for those cooperatives making direct foreign sales. A few managers said that the commercial banks watched their loans more closely than the Banks for Cooperatives, and the closer monitoring of their business gave the manager a greater sense of confidence.

The commercial banks will probably become more important as a capital source as more cooperatives find need for additional services and find their credit needs going beyond Banks for Cooperatives' lending limits. In fact, the existence of unused lines of credit with commercial banks appears to be partial evidence that cooperatives are beginning to take steps to assure themselves of adequate credit in the future.

Probably the biggest complaint expressed about the commercial banks was that their interest rates tended to be higher than the Banks for Cooperatives System. This may be especially true for smaller cooperatives or those in a less than sound financial position. A second concern was that commercial banks often did not really understand cooperative financing. This deficiency is improving as more cooperatives seek commercial bank lines of credit and explain their financing methods and needs. Currently, cooperatives are being sought by commercial banks and investment bankers that wish to become cooperative lenders.

Debt Securities. All the core sample cooperatives and many other cooperatives interviewed had some form of outstanding security unrelated to patronage refund and per unit capital retain allocations. Issuance of stock, a common form of cooperative paper 10–15 years ago, is being gradually replaced by debt securities. The trend has followed the drift away from Section 521 tax status. In these cooperatives, income tax must now be paid on stock dividends. Interest paid on debt instruments is deductible, thus making debt securities an attractive alternative to stock issues.

Core sample cooperatives indicated no further major plans to shift to the use of nonpatronage securities (both debt and equity).

Most smaller cooperatives will continue to acquire their capital needs through Banks for Cooperatives loans rather than from debt or equity securities. But the large regional cooperatives may find it desirable to find supplementary sources. Those large cooperatives not using nonpatronage paper may be overlooking the pool of capital available from cooperative membership and rural residents acquainted with the cooperative. A noncooperative company without superior bond ratings may find it difficult to raise new capital today, but a strong locally or regionally known cooperative may find it relatively easy.[105]

The issue of "going to the public" for capital is closely associated with the securities issue. The subject is controversial. One cooperative manager aptly pointed out the conflict often found on the issue between the cooperative board and management: "The management doesn't feel too frightened about the idea of going to the public, but the board is very nervous about this type of thing. The growers are just not sophisticated enough in the area."

Problems exist with going public. A farmer cooperative does not sell voting stock to the nonmember public without violating various federal and state cooperative statutes which limit membership to agricultural producers. Issuance of voting stock to the nonmember public would also affect compliance with the cooperative provisions of the Internal Revenue Code allowing tax deductions for patronage refunds. The provisions of the Capper-Volstead Act allowing farmers to act together might also be violated as well as the exemption from securities registration. So "going to the public" in the sense of giving control to nonmember interests is obviously an unacceptable alternative for cooperatives. It is unlikely that a farmer cooperative would really wish to do this anyway.

Cooperatives can sell nonvoting preferred stock as well as debt issues to the general public. Since most of these public debt issues are subordinated and noncollateralized, their issuance provides little risk from outside control. Yet, concern is justified, especially if the public capital markets supply a large percentage of cooperative capital needs. Bernard Schulte, past president of the Central Bank for Cooperatives, says, "If the cooperatives are going to use outside sources who aren't really aware of cooperative purposes and where we are going, then troubles compound when adversity arrives."[106] Nonmembers may have little or no interest in the survival of the cooperative. To them, the recovery of their debt or equity takes precedence over perpetuation of the cooperative.

Public capital sources should not be ruled out by cooperatives. They are sometimes available, and for those cooperatives with a good financial condition and good investor contacts, they may be a logical source.

Tax-Free Bonds.[107] Tax-free bonds (often called industrial development bonds) are a possible source of capital for a few cooperatives. Land O'Lakes used them as one of its major sources of long-term debt. These bonds are long-term obligations issued by a local government unit. They are usually issued for construction or improvement of plant facilities that are expected to bring tax revenue, employment opportunities, or other business benefits to the community. The interest income to bond holders is tax exempt, giving cooperatives the opportunity to finance facilities at rates as much as 2 percentage points below the Banks for Cooperatives rate. Tax-exempt status is limited by federal law to $5 million bond issues (except for pollution control bonds), making this the upper limit on the use of the method.

Cooperatives that have studied tax-free bonds consider them to be an excellent source of financing mainly because the bonds carry a low interest rate, long repayment terms, and are issued for 100 percent of the facility's cost. A recent study of nine cooperatives discovered that the nine had used over $93 million in tax-free bonds over the 1949–77 period.[108]

Bond issue costs are high for these bonds, and the $5 million limit on tax-free issues restricts the use to other than large-scale projects. Nevertheless, industrial development bonds should not be overlooked as a relatively inexpensive source of long-term financing for facility investment.

Leasing. Leasing is a source of capital, although it is actually a means for acquiring operating control of an asset rather than ownership of it. In the United States, leases represent a business of over $5 billion annually.

Ordinary leases have been used regularly by cooperatives, but the newest development in the leasing field is leveraged leasing. The impetus for this form of lease grew out of bank holding company legislation and Federal Reserve regulations passed in early 1970. Lyle Fensterstock, Salomon Brothers, Chicago, Illinois, was one of the early innovators adapting the idea of leveraged leasing to cooperative finance.

Few cooperatives are now using leveraged leasing, although many use ordinary leasing to some extent. Mississippi Chemical Corporation (an interregional cooperative) is an example of a cooperative making use of a leveraged leasing program.

Leveraged leasing has several *advantages* as a cooperative financing tool. It is a tax-oriented lease for the leasing company. The lessor can claim most of the tax benefits associated with the ownership of the property. These benefits consist of depreciation and the 10 percent Investment Tax Credit. A cooperative generally would not benefit directly from the tax benefits of asset ownership because it has relatively little taxable income against which a credit can be applied. The tax benefits derived by the lessor are passed through to the lessee in the form of lower lease payments. In the case of leveraged leasing, the lessor acquires ownership of the asset by putting up only a portion of the capital (20–40 percent). Of course, the leverage available to the lessor is his ability to take all tax benefits incidental to full ownership even though he provides only a portion of the capital. The remainder of the capital is borrowed from institutional investors on a nonrecourse basis.

The IRS and FCA regulations prohibit the Banks for Cooperatives from supplying the debt capital for cooperative leveraged leasing.

Leveraged leasing provides 100 percent financing of leasable items. The result is an improved cash flow and evening out of patronage refunds. Other advantages of leveraged leasing may accrue to a given cooperative.[109]

Leveraged leasing carries with it several potential disadvantages. First, the concept does not begin to be practical until the $3 to $5 million level. This rules out its use for many cooperatives. Second, the leasing arrangements are quite complicated and require experts in legal and financial areas, such as an investment banker or a leasing broker, who arranges for equity financing and registration of lease terms. Third, residual value, if the cooperative wishes to buy the asset at the end of the lease agreement, is a problem. Management of a large multi-regional cooperative stated that leveraged leasing may be appropriate only if the asset is relatively fixed in place such as a plant on cooperatively owned land.

Possibly the greatest disadvantage is the dependence on current tax legislation. Changes in this legislation could undermine tax advantages to the lessor and make it difficult to justify the long-term commitment required. IRS guidelines have set down tough limits on what constitutes a bona fide lease.[110] A lessor must now have at least 20 percent equity. When the lease expires the asset cannot be sold to the lessee at a price far below current market value. IRS must satisfy itself that the lease is really a lease, not a disguised sale to the lessee.

Since cooperatives do not have the same tax liability as proprietary firms, they cannot take full advantage of tax-related investment incentives legislated for other forms of business. The leveraged lease concept is one way of taking such advantage. Most managers interviewed were familiar with the leveraged leasing concept. The concept will continue to receive attention by growth-oriented cooperatives, provided tax legislation remains favorable. But such arrangements should be carefully analyzed.

REGISTRATION OF SECURITIES. Currently, those cooperatives that have given up their Section 521 tax status may be subject to registration with the SEC if they issue securities that come under regulations.[111] A non-"exempt" cooperative can legally issue securities without registration if it restricts sales within one state, restricts sales to less than $500,000 in any 12-month period, or if it restricts sales to a carefully defined member group.[112] But these exemptions from the registration requirements are not clearly defined with respect to cooperatives. The problem includes not only the initial offering but also the trading of paper after the offering. While a patronage refund or retain certificate has generally been considered exempt from registration initially, its subsequent use as banking collateral or its use as a marketable security may place it under special scrutiny of the SEC at the present time.

The important point is that when the cooperative begins to use nonpatronage refund paper as a source of capital, it should seriously consider the question of SEC registration. Jerome Weiss has said, "Most commentators seem to agree that because of the broad scope of the definition of a security given by the courts and recent interpretations given by the staff of the SEC, cooperatives cannot afford to conclude they do not offer or sell securities within the meaning of federal statutes." Weiss goes on to say, "All sales of securities by cooperatives (non-"exempt" cooperatives) unless made to a carefully defined group under special circumstances, *may* be considered public offerings subject to SEC regulation."[113]

Registration carries with it some serious disadvantages to cooperatives.

First, registration may initially cost as much as $125,000 for a large cooperative, with a significant cost associated with the annual update.[114] Second, much management time can be used in providing the necessary historical data and annual updates. In some cases, supplying the necessary historical data may be next to impossible. Third, a few cooperatives are concerned that disclosure of the required information might open up previously confidential data. Fourth, failure to disclose all relevant information in a registration statement may result in charges of fraud by a disgruntled member. Court action could then require that *all* previous security issues be brought up for review.

The requirement for *full* disclosure is a substantial task and carries with it substantial responsibility on the part of management. Of course, cooperatives do have a responsibility to keep their members fully informed regardless of registration needs.

Registration has some advantages for cooperatives. It allows cooperatives access to new capital markets. Noncooperative businesses can have direct access to capital markets with investment attractions that cooperatives cannot match, such as sale of a publicly traded stock capable of fluctuating in value. Cooperatives cannot do this nor can they provide the holder of their paper with any direct participation in company net margins (limited dividends are allowed) except to the extent the investor is also a patron. Also, the noncooperative firm is often better known to the investing public.

The cooperative, through its registration, presents itself to the investing public and is able to market nonpatronage securities to a broader group of investors. Security title can also be transferred among members or nonmembers. The present legal environment may be another justification for SEC registration, especially if by failure to register its paper the cooperative invites scrutiny by regulatory agencies.

As of April 1976 only the following cooperatives were registered with the SEC: Agway, Inc., and its noncooperative subsidiary, Curtice-Burns, Inc.; Farmland Industries, Inc., and its cooperative subsidiary Farmland Foods, Inc.; Mississippi Chemical Corp.; American Crystal Sugar Company; and National Grape Cooperative Association, Inc., a Section 521 status cooperative. National made the decision to register because its patronage-related paper is transferable and is being freely traded. Midland Cooperatives, Inc., and Pacific Supply Cooperative have registered in prior years.[115]

Cooperatives not qualifying for Section 521 status should consider the above advantages and disadvantages of registration if they intend to issue any paper not clearly exempt from registration requirements. Even the large Section 521 cooperative needs to consider these issues if it is considering relinquishing its Section 521 status or feels the status is in danger of being revoked, or if it intends to make its paper transferable. The current regulatory climate appears to be quite unsympathetic to any request for special cooperative exemptions. Little change in that environment is expected.

7 Putting Strategies to Work

A COOPERATIVE'S STRATEGY, no matter how perceptive, is useless until brought alive. Somehow the strategy must be formulated and implemented. Formal planning can be used to take the strategy from the drawing board and put it into the framework of everyday business. Planning enables the activities of the cooperative to be programmed around its strategy.

Planning is necessary because the resources available to the cooperative are limited. Moreover, doing anything takes time and time is a resource that cannot be wasted. Cooperative managers must make decisions that use available resources most effectively over time. This is done by making decisions with a sense of futurity about them, or "planning," as Peter Drucker would call it.[1]

Planning makes the most of uncertainty. In planning, as much knowledge as possible about future conditions is gathered and evaluated. Business strategy is then formulated in light of this knowledge. Hopefully the strategy selected will capitalize on favorable forces and counter negative ones. This approach to decision making implies cooperative growth according to Richard W. Schermerhorn's definition of growth as "successful adjustment of operations to business conditions."[2]

The importance of planning cannot be overemphasized. In the complex, fast-changing environment sketched earlier, planning has become a crucial factor. Formal planning is partially a defensive but is also an offensive weapon. In contrast to companies that do not plan, companies that plan tend to be more aggressive and better sellers of goods; they control margins so as to reap greater profits and earn higher returns on capital.[3]

Essentials for Planning

At least five general "musts" represent the necessary, although not sufficient, ingredients for a successful planning effort. They are (1) a formal planning system, (2) a set of complementary plans, (3) a suitable planning horizon, (4) full participation in the planning process, and (5) an effective board-management relationship. Neglect of any of these basic needs indicates a misunderstanding of the basics of planning, and action needs to be taken to correct the misperceptions.

A *formal planning system* coordinates and combines planning ''activity'' so as to produce a single document that articulates the cooperative's strategy. This is done by having a system that specifies the periodic planning tasks to be accomplished within the organization, and then through an iterative process assembles a corporate plan. Through the iterative process better knowledge of the planning of others, better awareness of the assumptions and forecasts of others, and better awareness of past actions and future plans enable managers in the cooperative to take advantage of business opportunities and avoid inherent pitfalls in the environment.

Specific benefits attributed to having a well-documented corporate plan are:

1. Decision makers are forced to think about long-term as well as short-term results in creating the plan.
2. A comprehensive corporate plan ensures a clear, definite course of action that improves decision making.
3. The responsibilities of each person and each department are clearly defined.

However, these benefits will be recognized only if the formal planning system is integrated with other management systems of the company, such as the management control or information system.

A *set of comprehensive plans* is the second essential. Decisions must be made at various levels in the business and involve varying degrees of futurity. Various types of plans are necessary to meet these different needs. They are best described as operational, project, strategic, and corporate plans.

Operational plans cover the array of activities that the firm will actually undertake. For this reason operational plans have a very short time horizon. Starting with goals and objectives for current operations, usually by division or commodity, detailed plans are developed for marketing, production, labor, and finance.

Project plans deal with projects the cooperative is appraising or to which it is committed. These plans are initiated to assess the impact of certain changes in current operations—either changing the scope of present operations or redirecting them. Thus project plans can be very diverse, ranging from those dealing with upgrading a trucking fleet to those dealing with the creation of a new division.

The strategic plan presents a broad course of action that is the optimal way of attaining the cooperative's mission given the environment—a strategy. It specifies the kind of cooperative the members want, the scope of the business, objectives of the business, goals to be attained, and ways to measure accomplishments. The strategy is described with reference to product-market posture, service to members, return on investment, cooperative size, and relationship with employees and external institutions.

The corporate or long-range plan is the overall plan that comes out of the coordination and combination of the above plans.[4,5] Such planning is more effective than individual planning activities. The dimensions of this activity that make it superior are:

1. Comprehensiveness. On a regular basis the business is reviewed as a whole in relation to its environment. The formal examination and analysis of strategic problems and opportunities help the business move aggressively into the future.

2. Structure. The integration of the different planning activities requires a formal and systematic process within the business to ensure that strategic, project, and operations planning are effectively carried out on a coordinated basis.

3. Analytical process. Through interaction of managerial personnel throughout the firm, managers are introduced to new ideas and techniques of information processing and analysis that assist in decision making.

The corporate plan is the most used plan for top management. It represents the end result of the whole planning process. It can be used by top management to get hold of the big picture and make comparisons among divisions.

A *suitable planning horizon* is the third essential for good planning. The planning horizon is the time frame in which an informed look ahead is being made. The time period used should be long enough that realistic and responsible planning can be performed. Specific needs should be anticipated far enough in advance so they can be filled in an orderly manner. Industry peculiarities, market demand, resource availability, product lead times, and strategic objectives of the cooperative are all factors that can affect the time horizon of planning.

Two basic criteria are used when determining the time framework for planning. The first is the accuracy of prediction: the business should look only as far into the future as can be estimated within reasonable probability using the best available information. The second criterion is the commitment of the management to the future: operating executives must be willing to commit time and thought to events that are distant and do not have a direct bearing on current operations.[6]

Corporate planning requires that the different types of plans have compatible time horizons so they can be integrated to provide an overall plan. Generally, corporate planning has a 3-, 5-, or 10-year horizon because this seems a reasonable compromise between the cost of detailed long-range planning and the need for an adequate time perspective.

The length of time for which planning has been practiced also affects the planning horizon. As the planning functions develop over time, the horizon of the corporate plan extends further into the future. This extension has been evidenced in noncooperative industry through a number of studies. A study of long-range planning in 1939 by Stanford University found that about one-half of 31 companies interviewed planned in some detail up to a year in advance.[7] Only 2 firms had plans for as long as five years. A 1956 survey by the National Industrial Conference Board found that 142 of 189 companies employed a planning program with more than a one-year horizon.[8] By 1967 *Business Management* found that 50 percent of 101 companies surveyed did planning for three or more years.[9] In 1973 the Planning Executives Institute reported that 86 percent of the 380 firms surveyed had planning horizons of three to five years.[10]

Full participation in the planning process is the fourth essential. Planning can be an effective tool only when every individual and group expected to contribute to the achievement of the plans is involved in their development. Only when those responsible for performance are committed to the plans will they stand a chance of being met.

Planning is necessarily a line function. Staff, however, plays a crucial role in the planning function, as will be seen later. Producing a planning document is a futile staff effort if the decision-making executives do not involve themselves in the formulation of plans; or if, having done so, they run off in different directions. Formalized planning should never be divorced from actual planning nor, even more importantly, the decision making. To do so only spells trouble.

An effective board-management relationship is the fifth essential. An effective relationship is needed so that both parties can work within the planning framework to develop a program of action that characterizes the strategy. The board-management team holds many of the keys to cooperative survival. The creativity rests here; the ability must rest here. The strategy cannot be developed and mobilized if the board and management cannot work together effectively.

The board of directors and management form a team that works together to formulate and implement strategies. Each member has its own tasks although the role of the two parties in this team has never been clearly delineated. Generally the board and management are expected to work together constructively—the board establishing policies in consultation with management and management taking action with board approval. The board should encourage the success of the cooperative, not just avert disaster. To this end, the board should raise issues, provide innovative thinking, and ask probing questions of management instead of just "rubber stamping" management proposals. Management should respond to board requests and utilize the maturity and experience of the board in making important decisions. The operating decisions, once made by management, should be supported by the board.

How to Plan

Planning is a dynamic process within any cooperative. The overall flow of planning information within the cooperative should be as follows:[11]

1. Headquarters establishes the mechanics of the process and the assumptions and premises to be used.

2. Operating units prepare plans in accordance with the desires of headquarters.

3. Headquarters consolidates, reviews, and evaluates the plans and then communicates its assessment of the plans.

4. Operating units adjust plans to account for the observations and suggestions of headquarters.

5. Headquarters consolidates, reviews, and passes the total plan to the board for approval.

Management and the board should not assume that there is a "proven ten-step" procedure to planning. Experience has shown that no single planning process will suit every business. Each cooperative must "tailor" a formal planning system so that it is appropriate to its requirements and structure. Information is obviously one of the first concerns in planning, whatever the planning system. Not only is the planning system dependent on information not already available but it also makes much use of existing information such as the accounting system that generates reports for purchasing, operations, marketing, and control. The development of a planning system should make the most of this and attempt to integrate the new system with those already in existence.

Although it is not appropriate here to be specific about the planning system, it is appropriate to discuss the role of the key participants in the planning process. They are the board of directors, management, the planner, and the economist.

The *board of directors* should play a major role in the conceptual and strategic stages of the planning process. Conceptualization of the plan is most effectively carried out by a small group: a special planning committee of the board,[12] the manager, and the planning coordinator. This group should review the climate, conditions, and opportunities of the business. The group should then select the general, organizational, and facilitating strategies that optimize the total productive accomplishments of the cooperative. This is the statement of strategy. The strategy statement should then be passed on to the appropriate managerial executives for criticism and revision. The strategic plan is finalized when the necessary changes have been made and approved by the small planning group. It can then go to the full board for approval.

The interaction between the board and management in the planning activity is invaluable. By working together, both are committed to the same objectives and policies. These policies and objectives can then be used by management to provide ultimate direction and assistance when generating and assessing project and operations plans.

The other major involvement of the board in planning should be a final review of the corporate plan at the end of the planning process. This ensures that the final plan meets the objectives set by the board.

Management has a more general role in planning. General managers (GMs) are the educators and evangelists of the planning effort. Moreover, a planning function is in their best interest. The major reason they are at the top is because of their judgment. Functional executives know the details of their jobs better than general managers. Planning is their best method for identifying, gathering, and organizing the information necessary to make wise decisions. Thus it is not surprising when researchers questioned chief executive officers (CEOs) in industry about the importance of planning they found it ranked high. A study by H. Steiglitz found that 65 percent of 280 CEOs questioned considered long-range planning their most important task and spent 44 percent of their time on it.[13] Another study of 470 presidents (of which 62 percent were CEOs in their organizations) found that planning occupied 34 percent of their time, followed next in priority by general administration which consumed 27 percent of their time.[14]

General managers set the whole tone for and role of the planning effort. If they do not set an example and demonstrate interest in planning, it will probably never be done. Senior line executives will let technicians do the planning and spend their time performing functions they see as more useful.

The potential for alienating senior management executives from the formalization of planning is high. One reason is that the executives' support may seem to declare that they have not been looking as far ahead as they should be. Also, they may not want planning formalized because they believe that once plans are recorded and disseminated they will have closed off their options and limited their "thinking."

The GM should get planning started by initiating the generation of a strategic plan that involves the highest levels of the management hierarchy. The strategic plan maps the overall direction for the cooperative.

Results of the strategic plan should be used to provide middle managers with guidelines as to the kinds of results expected. Middle managers then can generate operational (or divisional) plans based on these guidelines. Involvement of these managers in planning is important; they are the ones who are the closest to achieving the desired results. They are committed to their achievements only when they believe that the plans are achievable. The best way to ensure this is by having the middle managers develop the operational plans for their divisions.

When operational plans are generated they should be returned to the GMs for careful appraisal. Keeping in mind the strategic plan, they can compare operational project plans against the desired direction of the organization. The GM should make sure that operational plans pass both financial and company policy tests. Without a strategic plan, plans formulated by the divisions may be accepted as prepared by the divisions. This can lead to a corporate plan that fails to satisfy corporate objectives.

The final task of management is to combine the various plans of the business into a corporate plan and present it to the board for approval.

The *planner* (planning unit) has become essential in today's business as the planning function has become more pervasive and complex. Top management no longer holds sole responsibility for corporate planning. Many businesses now use staff specialists, planning departments, and outside consultants to assist in planning.

The planning unit within the organization plays an important role although its job is one of maintaining and facilitating planning rather than performing it. Specific functions of the planning unit are to:

1. Coordinate planning activities—the planner should organize meetings and prepare timetables for planning purposes. Deadlines must be set and work organized according to the necessary priorities so that plans are completed in the time available.

2. Arrange for collection of data—the planner, either in place of or in concert with the economist, should collect and maintain a data base of company performance and factors external to the firm.

3. Assist in planning—the planner cannot and should not plan for operating people who must assume responsibility for the results. But he can sell, stimulate, and indoctrinate line managers to their planning obligation and instruct them in techniques that can be used. He can also provide the premises and assumptions for and technical assistance in planning.

4. Put planning documents into meaningful form—the planner should consolidate all presentations developed in the planning process into a suitable planning document.

5. Provide an evaluative function—the planner should be capable of providing an evaluative function to review plans developed at various levels in the organization. The planner must be competent enough to provide the GM with a complete product that need not require the GM to understand the nuances of every decision and assumption behind the conclusion.

In private industry, a rule of thumb has been if the business has sales from $5 to $50 million, there should be a full-time planning officer.[15] In actual fact, the size of the planning staff is dependent on its roles. Planning units tend to be larger when the scope of the planning involves many activities (i.e., analyzing mergers, joint ventures, new ventures, and so forth); many organizational units are developing individual plans; limited planning expertise is available outside the planning unit; or responsibilities include nonplanning activities such as market research and economic analysis.

Economists are particularly valuable for forecasting what the uncontrollable, external influences on the business will be, and this appears to be their predominant function. Other contributions the economist can make include critically assessing economic information coming from outside the business; helping select goals to be followed and providing measures for gauging attainment; analyzing the capability of and resources available to the cooperative; and providing materials useful in evaluating performance of the enterprise—for example, information that quantifies the performance of the industry and the economy. Thus, although the role of the economist is not synonymous with that of planning (he does not provide a course of action for the business enterprise), he can contribute to the planning process in a number of ways.

What Are Cooperatives Doing?

Informal planning has always been an integral part of business management, but only in the decade of the fifties did business start to take an interest in formal planning systems. There was a proliferation of articles on the subject, but translation into company practice lagged by a few years. By 1964 a *Business Week* survey found 71 percent of 139 companies had formalized corporate planning activities. Leslie W. Rue reported that, by 1973, 86 percent of 225 companies surveyed had formal long-range planning processes.[16]

Most managers of cooperatives interviewed in this study came to the conclusion stated by one manager, "Most cooperatives are 5 or 6 years behind their non-cooperative competitors in planning." A study by the Planning Executives Institute (PEI) in 1973 found that of American companies that plan, 47 percent had been planning over five years, 33 percent between three and five years, and only 19 percent less than two years.[17] Of cooperatives in the core sample only 1 had been planning formally for over ten years, 1 for seven years, and the rest had started only in the last several years.

Cooperatives have been slow adopting formal planning because they have not felt the same pressures as noncooperatives. Cooperatives that started planning usually did so because they found themselves in situations where they were forced to. Capital constraints have sometimes provided the impetus for planning.

An example of this is CF Industries, Inc., Long Grove, Illinois. It believed that high capital requirements and resource needs of the fertilizer industry meant that a rational approach to the cooperative's growth needed to be taken. Member cooperatives were asked to sign ten-year contracts for their fertilizer needs. This in turn forced members to look at potential growth in their own fertilizer markets to estimate future needs.

Another example is Western Farmers Association, Seattle, Washington. It found itself starting to do formal planning because of serious financial difficulties. Initially, the planning function was directed at assisting operations, but the overwhelming financial considerations made the organization realize that capital restructuring, financial planning, and operations should be tied together.

In the future pressures that will increase the need for and use of planning are (1) restructuring of capital within cooperatives, (2) involvement in competitive consumer markets, and (3) capital needs exceeding traditional sources of supply. Another factor having a positive effect on the use of planning is an awareness on the part of new managerial personnel of the value of planning.

Managers in eight of the nine core sample cooperatives believed they

A meeting of the Board of Directors of Farmland Industries, Inc. Directors are elected by members of the cooperative; they hire the manager and establish operating policy for the cooperative.

had effective planning programs despite their late entry into formal planning. This did not mean there were no problems. Managers in the two cooperatives doing the most advanced planning were gravely concerned about the inadequacies of the management information system that supplied their planning information. They believed that the problem had been a long-run result of the conservative stance taken by earlier management. Both were aggressively trying to upgrade their accounting function, their information system, and management's ability to use the system. Only four of the core sample cooperatives appeared to have the broad scope of coordinated planning necessary to develop a corporate plan.

Most large cooperatives have operational plans; some call them budgets and others plans. Other cooperatives tend to do budgeting that merely extends current financial statements into the future with little regard to marketing and production plans. In some cases even pro forma budgets are poorly done.

Many cooperatives undertake project planning when they have projects of major importance. Some cooperatives also do project planning by division as needs dictate. Farmland Industries, Inc., Kansas City, Missouri, is a cooperative that does extensive project planning. Each year, two types of

planning groups are set up—one type focusing on divisional and commodity lines and the other on specific problem areas. Each group is then responsible for generating a ten-year plan that covers market strategy, policy, yearly capital investment, labor, and sales.

The main differences among cooperatives that develop project plans are the kind and number of projects scrutinized. Some cooperatives appeared to have done only one project plan in the last ten years while others developed a number of project plans every year.

Four cooperatives went so far as to quantify objectives and determine resources needed for several years. Assuming that the chief executives compare plans against organizational aims, these organizations do strategic planning.

The time horizon of the plans is an issue closely related to the type of plans being formulated. Most large regional cooperatives surveyed had planning horizons of four or five years. Smaller regional cooperatives had plans with horizons of two to three years. Few cooperatives had plans going over six years into the future. Many cooperative managers expect to extend the planning horizon of their business to either five or ten years. Several managers confused the time horizon of project plans with that of the corporate plan. They assumed that, because particular project plans were made for a certain number of years, the cooperative's total planning effort considered the same time horizon. This typically was not the case. Project plans had time horizons far beyond those of other types of plans.

A survey of cooperative directors in twelve cooperatives in this study showed that directors within individual cooperatives disagreed on the planning horizon of their cooperative.[18] Moreover, discrepancies between the time horizon given by management and that given by the directors are sometimes large. This indicates that the boards are either misinformed about what their organization is doing or they do not understand the different types of planning that can be done.

Those having a part to play in the formulation of plans are the board, management, the planner, and the economist. Managers in all core sample cooperatives told the researchers that the board set business policies, an integral part of strategy selection. Only one cooperative, however, had a board committee that developed and reviewed the strategic plan. Managers of two of the nine core sample cooperatives thought that the board should concern itself only with broad policies while operating policies should lie in management's realm. Management of three other cooperatives generated policy recommendations which were then sent to the board for action. In all three of these cases, cooperative directors believed that they still maintained effective control over their organizations.

All core sample cooperatives presented their annual budgets to the board for approval. Only in one case did a cooperative explicitly state that it presented a corporate plan to the board for approval.

The diffuse nature of planning throughout management did not permit a comprehensive examination of management's role in planning. Even so, the researchers thought that the time top management felt it should spend planning was indicative of its commitment to it. Managers in nine core sample cooperatives were asked if top management should commit a lot of its time to planning. Management of the three most aggressive and growth-oriented cooperatives strongly agreed that this should be the case. Three others agreed and the remaining three disagreed. Looking at the last three: one cooperative did no formal planning, another had recently gone through a number of major crises and was more concerned with correcting past mistakes than looking toward the future, and the last one viewed its environment as holding no major challenges for the business in the future.

The chief technique for linking corporate objectives with individual managerial peformance is management by objectives (MBO). Four of the core sample cooperatives had MBO programs. One of the four had found that its program had only mixed success. It seemed to work well when both a manager and his superior found it effective, but it had failed to work in several places because one or the other did not believe in it. The survey also found a tendency for organizations with MBO programs to feel that middle managers should play a larger role in planning than those organizations without such programs. This was fully expected in light of the purpose of MBO.

Several cooperatives have set up separate planning units, usually one person, specifically responsible for planning activities (FS Services, Agway, Land O'Lakes, Tri Valley, Gold Kist). Other cooperatives have tied part of the planning activity either to the financial or economics personnel (Farmland, Western Farmers). The typical role of these units is scheduling, planning activity, providing technical assistance in planning, and consolidating plans. Top management has maintained sole responsibility for evaluating the plans generated.

Cooperatives also have made use of consultants in planning. They have proved useful because of procedural, functional, and systems know-how—in part due to their experience with formal planning arrangements in other companies. They may be used on a continuing basis to provide insights into new thinking and techniques of planning.

The use of economists in assisting with planning appears to be very low in the cooperatives studied. Although five of the core sample cooperatives listed economic factors as having the greatest influence on the organization at present, only one of these cooperatives had any economists with an advanced degree. None of them had any intention of hiring any more economists. Other cooperatives with economists ranked economic issues as being of low importance in three of four cases.

Some Lessons Learned

Changes are necessary in many cooperatives if planning is to be improved. These changes stem from lessons learned from the cooperatives studied and from the experiences of noncooperative organizations.

Greater formalization of the planning system is needed. Formal planning systems in most cooperatives appeared rudimentary. Procedural rules and timetables need to be developed and formalized in procedural manuals for planning. Premises and assumptions used in developing plans should be written down. Computer planning models need to be developed to take the drudgery out of planning and permit sensitivity analysis of plans. These efforts will provide an improved framework for the planning effort that will ultimately lead to better plans.

The quality of planning also needs to be improved. Few cooperatives develop comprehensive plans. Much of what they do is disjointed. Groups within the cooperative work independently developing budgets and plans and never get together to make certain the results are compatible—they are both aimed at helping move the cooperative in a particular direction. Moreover, few cooperatives do as much planning as they should. A complete and comprehensive set of plans will ensure that a logically consistent approach is being taken to the fulfillment of the cooperative's objectives.

Better delineation of the planning task is another necessary change. For example, directors and management have different roles in planning. Different members in the management team have different roles. Proper delineation of tasks will guide and constrain the activities of each in planning. This will eliminate redundancies in the work load and minimize conflict between the board and management and among members of the management team.

The manager must demonstrate his interest in planning and set an example. Planning will probably never be done properly if he does not do this. Senior line executives will let technicians do the planning while they spend their time performing functions they deem more useful.

Middle management must be actively involved in planning. Management by objectives is a managerial philosophy that may promote this involvement.[19] Unfortunately, it has had only mixed success in several cooperatives that have tried it. The program failed to work because some managers in the cooperatives did not believe in it. Some acceptable method is needed to motivate all managers to achieve goals that contribute to the business.

The board of directors must undergo a number of changes before it can work competently in the strategic process and maintain a productive dialogue with management. The members of the board should be better informed about their role in the planning process. Many boards need either to have the number of directors reduced or to subdivide it into subgroups or committees for handling detailed business. And members of the board should be adequately rewarded for their efforts. Improved compensation will increase the number of persons attracted to the director's job beyond those who come because of their loyalty to the cooperative and a desire for enhanced community respect.

To achieve its mission a cooperative must select a strategy that considers cooperative growth and adaptation to a changing environment. Aspects of the strategy should be articulated in organizational plans. Every individual and group expected to contribute to the achievement of the plans should be involved in the planning process.

Experience has shown that no single planning process will suit every company. Each company must develop a process most appropriate to its requirements. However, cooperatives should formalize the process because it facilitates decision making. A formal planning system is nothing more than a set of policies and procedures designed to improve management's decisions. A formal planning system makes management's decision making deliberate, comprehensive, and structured.

Cooperatives have had a late start in developing formal planning systems vis-à-vis their noncooperative competitors. In part, they have not felt the same pressures to plan that others have. As a result, cooperatives are lacking planning expertise, planning models, and adequate information systems required to backstop detailed planning.

While much of the cooperative planning is still rudimentary, most managers in cooperatives hope to articulate their business strategy in a corporate plan in the near future. The feasibility of this depends on their ability to upgrade the planning functions within the cooperative. This will require investment in people, systems, and information. It will also require that the board and management work effectively as a team. In many cases changes are necessary in the board. Moreover, management needs to be encouraged in the planning task.

Cooperatives are moving in this direction now, but planning and control need more emphasis. Improved strategies necessary for cooperative survival in the next decade will do little good until plans to put them into action are made and followed.

The Strategy Team

The board of directors and management form a team that must keep the cooperative in step with the changing environment. The major function of this team rests in formulating and implementing strategies and then monitoring and controlling the organization to make sure the strategies are followed.

Response to the forces affecting the strategy of a cooperative calls for integrated and effective teamwork. Compromise between the team members is the key to this. It is necessary because of the unique structure of cooperatives. Management likes to consider the cooperative an individual, self-sustaining, economic entity because this facilitates the use of traditional managerial procedures for evaluating performance and maintaining control. Directors, on the other hand, view the cooperative as an extension of the patrons' farming enterprises that provides social as well as economic benefits. Unlike noncooperative directors who view owner interests as being financial for the most part, cooperative directors are nearly always engaged in farming. Thus they equate owner interests with particular products and services. This means that the cooperative must reconcile differences between the board, which is concerned with progressiveness, and the management, which is concerned with efficiency of operations. Noncooperatives, on the other hand, can have conflicting objectives and still come out all right because the two sides—owners (stockholders) and patrons (customers)—can be dealt with independently.

The unique structure of cooperatives provides the cooperative with certain advantages. The owner-patron relationship of board members means that those responsible for guiding the cooperative have intimate knowledge of what both owners and patrons want. It also provides special legal treatment. For example, joint efforts of noncooperatives are in violation of the trust laws; however, cooperatives are able to work together whenever it serves their best interests.

Cooperatives make a unique contribution to the market economy. The cooperative form of organization has allowed farmers to undertake collective action in certain areas such as marketing while protecting individual farmer decision making regarding production. Thus farmers are provided a means of countervailing the power of corporate agribusiness while maintaining their individualism. Cooperatives are a mechanism that provides for a competitive market both within agricultural production and with other sectors of the agricultural and food economy.

The opportunity for modern cooperatives lies in the ability they develop to reach a compromise between a management that seeks to innovate and a board that seeks to protect the interests of owners and patrons.

Compromise will be most successful when members of both teams recognize their tasks and their abilities and then work constructively with one another. The nature of the ultimate compromise will depend largely on the abilities of each member.

MANAGEMENT

Managerial Capability. Historically, the survival and success of many cooperatives have been attributed to the ability and leadership of a strong manager. Autocratic rule is often needed in the early life of an organization; but an organization as complex as a modern regional cooperative may fail if it relies on the ability of one man to manage it. Managerial dominance can also lead to a poor distribution of skills within top management since managerial skills are often not developed because of lack of opportunity.

To assess the current breadth of managerial ability each core sample cooperative manager was asked if he believed the cooperative could survive if it lost a number of top executives. All cooperatives in the core sample believed that they could survive if they lost their two top executives. If the five top executives were lost, one cooperative did not believe it could survive, two did not know if they could survive, two believed that it would have a serious impact but that the cooperative could survive, and the remaining four cooperatives did not believe that it would affect survival potential. The implication of this assessment is that these managers believed that they had broadened their managerial capabilities and had a broader base of managerial competence than indicated by earlier literature.

Agway, Inc., a multi-regional cooperative headquartered in New York State, has developed a continuing system for dealing with the possible loss of top management. It practices what it calls "successive management." Every six months the top 5 executives in the organization review the top 32 management positions in the cooperative to see who could fill the positions if a replacement were needed. They also look at the next 200 men in the cooperative to see who could be moved to another job if necessary.

The authors were told by a cooperative manager in the Midwest, "I am impressed with the quality of other cooperative managers; but, when I look within our organization I'm very worried about the quality of management at the third and fourth levels." This problem is evident in many cooperatives and occurs because of the nature of cooperative management. Stability in cooperative management has tended to discourage more aggressive individuals from joining cooperatives. A number of cooperatives try to hire personnel with backgrounds in farming. Such a focus may be justifiable when a person is going to be the manager of a local cooperative but is questionable for a staff position in headquarters. These attitudes have restricted cooperative ability to hire management types.

Some progressive cooperatives are not adverse to going outside their organization to get necessary managerial talent for major staff positions.

One manager said, "Members don't seem to mind if people are not brought up through the ranks provided they know the job." On the other hand, managers in less progressive cooperatives "feel strongly that people in top management must be brought in at the lower levels and trained for top management positions. In-house promotion builds good morale and makes sure that the man knows the organization."

A number of cooperatives—Indiana Farm Bureau Cooperative Association (IFBCA), MFC Services (MFC), and FS Services, Inc. (FS), in particular—have tried to upgrade middle management by providing capable managers for local cooperatives. Traditionally, many local cooperatives hired and paid their own managers. Often the farmer-directors involved were not aware of the skills, training, and knowledge a person needed to be a good manager in a local cooperative. They tended to pay low salaries and thus hired low quality managers.

IFBCA started providing managers for some local cooperatives in which financial problems existed. Managers were hired, placed on IFBCA's payroll, and the locals charged a management fee. Most of the locals participating in the program have liked the arrangement and want it continued.

MFC began a management training program about six years ago when top executives were displeased with the performance of local managers. Many local managers are now hired by MFC and go on MFC's payroll. These managers usually have a college degree and go through a two-year training program.

FS Services, Inc., brings about 50 to 70 college graduates into its system every year. Membership agreements provide FS with the authority to approve managers of some of its affiliated local cooperatives. FS has set up a labor pool, and when a local cooperative has a vacancy, FS sets up an interview and provides candidates to be interviewed. FS management thinks that its program has helped to strengthen the locals by strengthening managerial control at local operations.

In general, regional managers thought that management expertise at both the local and regional levels has greatly improved over the past several years.

Managerial Training. The authors were told by management teams in many cooperatives that management was "up-do-date and quite capable." However, the evidence indicates that cooperatives have been slow in upgrading the abilities of older personnel and hiring new personnel trained in the business skills and quantitative methods of fundamental importance in the anticipated environment.

Fortunately, the more progressive cooperatives provide training programs for managerial staff. A good example is Land O'Lakes, which has undertaken a 16-hour, intensive training program in financial management for its top and middle management. It has found that managerial personnel are receptive to training courses and eager to learn.

Land O'Lakes also takes its professional accountants to a retreat for two days a year to review what has been happening in accounting. Some cooperatives also bring in consultants to provide training courses or encourage their people to attend schools and seminars.

The importance of training in developing managerial capability cannot be overstressed. Hopefully, few cooperative managers have the philosophy about training that one manager had: "We have no management training program. On-the-job experience is the preferred method due to the cost of providing the training for personnel."

Managerial Compensation. Compensation is a major incentive for improved performance in any company. Cooperative boards are learning that they must pay higher salaries than in the past if they want to attract and hold top-notch people.

Managing the progressive cooperative is no easy task. Such a cooperative needs a knowledgeable and decisive management team that is held together by a credible reward system. In the past, cooperative compensation has lagged behind that of noncooperative businesses. This may be due to the way in which top management has been recruited. Many top executives in cooperatives have had to work their way to the top through promotion within their own organizations. Thus cooperatives have faced less pressure from outside for increasing executive salaries. Inbreeding of management within the cooperative has been encouraged. Moreover, many good executives have not looked to cooperatives for jobs because of lower rates of compensation.

Comparing compensation paid cooperative personnel with that paid noncooperative personnel is difficult because most companies like to keep their compensation programs confidential. Some feeling of the difference in compensation can be acquired by comparing the compensation paid the CEOs in different companies. Compensation of the CEO sets an upper limit on compensation and therefore can be used to evaluate the size of the gap in compensation among companies.

Average compensation paid cooperatives' CEOs is still behind that paid CEOs in other corporations. In a study prepared for the National Council of Farmer Cooperatives on CEOs' compensation, it was found that "on the average, the cooperative chief executive receives about 70 percent of the income as that of his counterpart in corporations." In 1975 the annual average salaries of cooperative managers ranged from $35,000 for cooperatives with sales of less than $25 million to $100,000 for cooperatives with more than $500 million in sales.[20] The average annual salary for a CEO in corporations with sales in the $200–500 million range was $122,500, according to a survey by the American Management Association.

The major element in compensation packages is salary. Salaries vary considerably from one industry to another and one area to another. They also vary according to the bonus received. The base salary for those receiving bonuses are much lower on the average than for those not receiving bonuses; however, recipients of bonuses usually have higher total salaries than nonrecipients.

Bonus and incentive plans were used by 38 percent of the cooperatives surveyed in 19 percent of the executive's total compensation. Seventy-four percent of the noncooperatives used these plans for providing 40 percent of total compensation. Obviously, bonuses play a far greater role in noncooperative than in the cooperative compensation. In part, this is a result of the nature of cooperation. Many cooperative bonus programs are more a "sharing of the wealth" than true incentive programs. One cooperative principle is operating at cost. If managers are given high bonuses, they are increasing costs to some extent. Hence, many boards like to keep bonuses as small as possible.

About 33 percent of the cooperatives and 36 percent of the noncooperatives had deferred compensation programs in 1973. These programs were arrangements providing company payments to executives after retirement or termination of employment through no fault of their own. The bulk of the plans used by cooperatives provided income through bonuses, savings plans, and contractual arrangements.

Other perquisites provided in nearly all companies were life insurance, health insurance, and pensions. One benefit not offered by cooperatives and of major importance in other businesses was stock options. About 81 percent of the industrial corporations provided their executives with stock options. Cooperative principles and practices, on the other hand, do not allow managers to own voting stock in their cooperatives.

Over the last ten years the gap between cooperative and noncooperative compensation appears to have narrowed. Some managers of progressive cooperatives are now using data such as that of the American Management Association's Executive Compensation Service to determine competitive levels of managerial compensation.

Some managers still think the difference in compensation between cooperatives and noncooperatives is not important. One manager said, "While compensation paid to top cooperative management is substantially less than that paid to top management in the private sector, very few cooperatives have gone broke because of poor management. A lot of private companies have because they lacked good managers." This point may well be challenged on the face of it, and it raises a corollary issue, Can cooperative managers afford to be as conservative in the future as they have been in the past? Not taking a few risks may lead to a more stable record of performance, but it might not be in the best interest of the member-patrons.

A Future in Cooperative Management. Many cooperative managers are excited about the future for young people choosing a career in cooperative management. It can present a unique experience. Working for a cooperative can have substantial benefits. In many noncooperatives, said one manager, "the investors are only interested in a return on investment. They do not give a darn about developing people, services, products or anything else. They just want to know how much interest they are making on their money. It's no fun working with that type of people." Working for farmers is a pleasure. They "are honest; they have been through the wringer, they have had adversity, they are willing to go with the punches, and they don't panic when there's a little trouble. They appreciate a good job."

BOARD OF DIRECTORS. The cooperative board that decides to formulate company strategy is a unique position vis-à-vis the noncooperative board. As stated earlier, directors on the cooperative board are intimately attuned to the needs and wants of patrons because they themselves are patrons. They have been chosen by the members and patrons so they know what the users expect in terms of service and performance. However, many boards need to examine their capabilities and resources before they formulate strategies.

Directorial Capability. Any assessment of the capability of the board of directors naturally centers on a discussion of three elements: the balance of directorial ability, the size of the board, and the board's knowledge. A well-balanced board requires a proper mixture of directors with differing experiences and skills. Cooperative members may be more interested in voting for directors who will represent their constituency than those who will balance and complement each other so as to create an effective board. More effective boards come about when members are educated as to what is needed on the board of directors. Leon Garoyan and Paul O. Mohn surveyed in detail the boards of three different types of cooperatives and found "weaknesses . . . in areas of broad economic sophistication, participation in cooperative trade organizations, and participation in local and regional organizations that have a special economic impact on their cooperatives. They are stronger than one would expect in membership contacts, but perhaps this reflects the politics of getting elected, and satisfaction from participation in membership meetings. We are not surprised by the high ratings shown by directors in product competence, obviously a strength of cooperatives."[21]

The contributions made by an individual director should be a matter for measurement and evaluation. The ideal director will meet the following performance criteria:

1. Committed and interested in the success of the cooperative
2. Available for cooperative meetings and functions
3. Briefs himself before meetings
4. Is a capable counselor
5. Has suggestions and observations that are valuable to management

Many studies have looked at the size of the board of directors. The Conference Board reported that the median size of a corporate board was eleven for manufacturing firms and thirteen for nonmanufacturing firms.[22] Another study reported a mean size of twelve with a range from three to twenty-six directors.[23] Still another study reported that 20 percent of all companies had five to nine directors, 40 percent ten to fourteen, 21 percent fifteen to nineteen, and 18 percent twenty or more.[24]

In one study, the average number of directors on a cooperative board was thirteen, with 90 percent of all cooperatives having fifteen or fewer. For those with over $20 million in sales, 3 percent had five or fewer directors, 32 percent six to nine, 33 percent ten to fifteen, and 32 percent fifteen or more.[25] While direct comparison with other companies indicates a similar size distribution, this result is somewhat misleading since the manufacturing companies surveyed above were considerably larger.

A major reason for large cooperative boards appears to be the geographical and commodity representation of board members. Regional cooperatives have often developed through merger of small cooperatives. Members have been unwilling to lose their representation on the board as cooperatives merged. Thus boards have been consolidated over time and have grown continually larger.

The number of board members should be determined by using two criteria—workable size and adequate representation of major viewpoints. When a board is too small it may not have enough members to do its job properly and may not represent all major viewpoints. When the board becomes larger it becomes cumbersome and unwieldy. Cooperatives with large boards usually turn to the use of subgroups or committees.[26] However, the effectiveness of the board in setting policies, providing counsel, and providing advice is determined more by its competence than by its size.

The board of directors, even though effectively structured, must be adequately informed if it is to offer sound advice and counsel to management. Boards are increasingly dependent on managerial staff for necessary information as businesses grow in size and complexity. Moreover, information systems are specifically tailored to channel information to the managers' offices for their own use. If the board is to be kept fully informed, management must work with the board to determine what the board needs to know, when it needs to know it, and how it wants the information presented.

A board likes to receive a number of information pieces prior to any meeting. Eighty-eight percent of the larger industrial companies provide their directors with information prior to board meetings. By far the most common information provided was financial data; 81 percent of the companies provided these data in advance while 21 percent provided marketing data, and another 17 percent provided manufacturing data. Twenty percent provided other materials such as minutes of the last meeting, and only 5 percent of the businesses provided an agenda in advance.[27]

For cooperative directors, unlike their noncooperative counterparts, the agenda is the most common information sent them. Of directors in cooperatives with sales over $20 million, 12 percent received the agenda in advance, 9 percent the minutes of the last meeting, 6 percent special information for decisions, and only 1 percent financial statements.[28]

These differences between cooperatives and noncooperatives are not directly comparable because the information on industrial firms was supplied by management on an organizational basis and the cooperative information was collected from directors. Nonetheless, it suggests that cooperatives have been less generous in supplying directors with information prior to board meetings. An effective board is impossible if it does not receive adequate information.

Directorial Training. In most cases, a new director lacks the expertise necessary to handle the functions required of a director. The result is that the older board members must see that each new director is "educated" and understands his or her responsibilities. Some managers and other cooperative leaders have recognized this need and, overall, have undertaken some excellent training programs. New directors have been given training and indoctrination sessions covering the scope and operations of the cooperative; distinctions between board and management responsibilities; special topics such as long-range plans, contracts, leases, bylaws, and financial information; and sources of further information on special topics.

Directors of many of the core sample cooperatives have taken either orientation programs for new directors or specific director training courses, typically lasting a day or longer. All orientation programs and most training courses are offered by the cooperative itself.[29]

However, the researchers found some cooperatives limiting training to those elements they deem essential. Few managers desire "rubber-stamp" boards, yet limited education may present a biased view of problems and lead directors toward conclusions that are not correct. Effective boards must be adequately trained if they are to be initiators of action rather than judges voting on management proposals.

Directorial Compensation. Compensation of directors varies substantially by organization, depending on the number of committee meetings, the size of the meeting fee, the size of the annual retainer, and the company's philosophy. The amount of director involvement also affects the level of compensation; directors more deeply involved in company affairs can receive compensation greatly above average.

Compensation paid by most businesses is a regular meeting fee—either a set fee for each meeting or a per diem rate—and/or an annual honorarium or retainer. Some businesses also allow outside directors discounts on company products or participation in other employee benefits. Apart from direct compensation, many businesses also reimburse directors for expenses incurred in attending meetings.

Studies of noncooperatives showed that the most common type of compensation given board members was a regular and committee meeting fee plus an annual retainer.[30] The usual fee per meeting ranged from $100 to $250, and the typical annual retainer was about $2,500.[31]

Most cooperatives pay directors a straight fee for each board meeting plus a travel allowance.[32] Compensation of cooperative directors does not compare favorably with that paid to directors of noncooperative organizations. It is unlikely that people seek cooperative directorships for monetary reward because compensation is seldom adequate to cover the costs of being away from the director's farm or place of business.

Assessment

The board of directors and management form a team that works together to formulate and implement strategies. Within cooperatives this team has unique advantages that are attributable to the cooperative form of organization that has special social treatment. These advantages enable the cooperative to be better aware of changing patron needs and to provide it with more ways of satisfying these needs.

Compromise within this team is essential if the cooperative is to make the most of its advantages. The satisfaction of patrons' needs must be balanced with owners' needs. This makes managing a cooperative a greater challenge than managing another type of organization because in the latter case these needs can be dealt with independently.

Reaching a compromise that results in an effective strategy is not easily accomplished. It requires perceptive individuals who realize that working toward a compromise of objectives will provide the best solution for the cooperative. Such people are not readily available. Often they must be trained and, once trained, must be adequately compensated for use of their abilities.

The authors found, in general, that today's cooperatives have more perceptive individuals on the board-management team than before. This has been a result of recruiting more capable individuals, giving training programs that have enhanced the abilities of these individuals, and providing improved compensation to encourage them to use their abilities to the greatest extent possible for the cooperative cause.

Cooperatives must not stop striving for further improvements in the board-management team. Cooperatives have performed well in recent years in narrowing the gaps between the quality of management in cooperatives and noncooperatives. Failure to continue this improvement will mean a widening gap between the two because noncooperatives will continue to progress in this area.

8 Cooperatives in the Future

THIS BOOK attempts to answer a question seriously asked in recent years—Will United States agricultural cooperatives survive another decade? The answer is yes. How well they will survive depends on many things, most of which are under the control of cooperatives. Thus they can control their own destiny. This book deals with how they can do this. Therein lies its value.

Changes in the environment are going on at an accelerated rate. Cooperatives were sanctioned to aid farmers in adjusting to such changes. How well have they done this? This book says fairly well.

Adjustments to change will be many and diverse. Changes in the environment for cooperatives may even change the purposes of many cooperatives.

Improved strategies and plans must be designed. New strategies for cooperatives are evolving. New planning techniques are being tried. The whole process is more formal and requires a more comprehensive knowledge base than in the past. Both management and board must be tuned into and be capable of handling this new era for cooperatives.

Planning processes recently have been improved and are being more widely used in American businesses. Several progressive cooperatives studied were experimenting with and adapting these new planning processes to cooperative needs. But most cooperatives were lagging behind noncooperative businesses in depth of planning.

Knowledge needed by cooperatives is broad and often unique. It comes from highly diversified sources such as other cooperatives, government agencies such as the ESCS in the USDA, the general business community, educational institutions, and both public and private research establishments. Such knowledge is often lacking or incomplete and is seldom specific to cooperatives.

The Environment over the Next Decade

The *scientific and technological environment* for cooperatives will find much more food and nutrition research being done by a larger group of government agencies and universities as well as by businesses generally. The research will be more worldwide in orientation; more crop oriented as compared to animal oriented; more nonfarm food oriented; more basic; and more focused on nutrition, health, and government regulations than currently. Cooperatives studied were gearing up with several different approaches to get this research from others or to do it themselves. But they are lagging behind noncooperatives and will find that they need to do much more to get on top of this area.

The *public policy environment* for farmers will place the burden of the minority role of agriculture more squarely on cooperatives. Importance of food in total public policy will be better established than during the last thirty years. Despite this, farmers through their cooperatives and other farm organizations will have a great need to cooperate among themselves. Cooperative leaders must stay aware of changes in commodity programs, marketing programs, the marketing system itself, and world food policy. Antitrust will be a key cooperative policy issue, and cooperatives will be under increased public pressures. Price and income policies of the next decade will have more production and marketing controls than during the last few years, but policies will not revert to the pattern of some historical ones emphasizing high price supports, restricted output, and low export volumes. More emphasis for the self-discipline of agriculture will be placed on farmers working through their cooperatives. Greater emphasis will be placed on commodity-oriented programs.

The *social environment* will continue to bring up new and powerful issues. Consumers will press for more input into agricultural issues. Increasing environmental and public concerns will pressure agriculture to stand on its own feet as an industry.

The *economic environment* for cooperatives will be at least as favorable as in the last two decades. Average farm incomes will be favorable relative to those of the last two decades but will be more variable. Prices of products sold and prices of purchased inputs will vary widely. Inflation will be a continuing and important problem. World markets will have a greater effect on cooperatives. The role of agricultural exports in farm policy will be of special significance. Cooperative patrons will demand a wider range of more integrated services and products.

The *competitive environment* will find cooperatives neither dominating the competition nor being dominated by it. New ways of organizing to do business will evolve such as multi-cooperative organizations, foreign subsidiaries, government-encouraged cooperatives, and specialized consulting and service businesses. Cooperatives will be more involved in more products, more services, and wider market areas.

In the supply and inputs field, cooperatives will need to decide where they can be big enough to stay competitive. This will raise questions of how far to integrate and how far to diversify. New multi-cooperative organizations will be organized; farmers will need to decide when their cooperatives should go it alone and when they should join others. More nonfarm supply customers will be available for the cooperative that wants that business. Availability of both raw material and finished supplies will be of great concern.

In the handling of raw farm products the real issue is whether the open, competitive market can continue to exist. Efficient handling of farm products will be stressed, and many cooperatives will consider whether to add other product lines in order to spread overhead and to assure market outlets. New marketing techniques, such as the teleauction and broader information systems, will be under experimentation. Cooperatives must give more leadership in such areas.

Consumer affluence will continue to stimulate innovations in the packaging and distribution of products and services. Energy shortages, expanded world food needs, and desire to avoid inflated food prices will necessitate cost saving and efficiency. The place of cooperatives in fulfilling consumer needs while yielding equitable returns to farmers will be a difficult but important one.

Changes in food distribution will hold the key to who will have access to the ultimate consumer. A big issue for cooperatives will be whether they must enter this competitive environment to protect their outlets. The key institutions with entree to consumers probably will still be the retailer and the national diversified manufacturer with strong national brands. But the newer, rapidly developing food service industry may be the most important one within the next decade. A few cooperatives have already entered the food service industry, but there are many others that are reaching the point where they must decide whether to enter this industry.

Cooperative Purpose

The *purpose* of a cooperative in such an environment is complicated. Historically, the basic purpose of a cooperative was to enhance the economic welfare of its patron-owners. This will remain the basic purpose, but how to do this will be more difficult. The adjustment processes and ways to fulfill a purpose have increased. Over time cooperative purposes have changed and they will continue to change.

For example, the purposes of the cooperative, or at least the means to obtain these purposes, are more like those of other businesses, more involved with marketing activities, more concerned about employee welfare, and more related to production input quality. Ownership and control by members is no longer taken for granted, and more thought is given to coordination of activities within a given organization and among organizations. Financial and management requirements are much greater.

The cooperative has a unique purpose—to provide farmers a means to work together to help themselves economically and socially. When cooperatives originated, their purposes were only one or a few. And they usually pursued their purposes individually. Today, most cooperatives have varying purposes, and increasing emphasis is put on accomplishing these purposes through joint organizations with other cooperatives. Such organizations will be expanded by farmers. However, farmers, directors, and managers will need to remember that the purposes of their cooperatives should not be taken for granted, and that they should be reevaluated often.

Strategy

Strategy is a broad course of action selected to achieve the purposes of the cooperative. Each cooperative must have this defined before it can fulfill its purposes. In many ways the strategy is where the action starts. Here change in the cooperative meets change in the environment. Farmers must recognize that their cooperative cannot avoid having a strategy; lack of a strategy unfortunately is a strategy.

Growth has probably dominated the strategy of cooperatives historically. Growth can and has covered a multitude of management errors. Cooperatives have potential for continued growth, possibly at an accelerated rate. Both managers and academic people who were interviewed thought that cooperatives must grow to survive.

Most agribusiness growth has been by external means. External growth is becoming increasingly difficult for large noncooperatives, and cooperatives are under public scrutiny about such growth. Most managers hope to continue to grow externally to accomplish their strategy. They will continue to test the legal bounds for this means of growth. External growth usually is easier than internal growth, but an emphasis on internal growth will be necessary for cooperatives in the future. Such growth will take thoughtful strategy and effective planning.

Growth probably will not be an adequate strategy for most cooperatives and it may also become more difficult. Possible limitations of growth will include (1) resistance by board and management; (2) capital constraints; (3) competition by other cooperatives; (4) competition by noncooperatives; and (5) the external environmental factors influenced by consumers, public attitudes, and regulations.

Thus cooperatives must also turn to other strategies. Primarily these will be strategies that allow for adjustment to better meet the changing needs of larger, more sophisticated farmers in the next decade.

Managers and others interviewed for this book thought that the survival of farmers and their cooperatives will lie mainly in the efficacy of three general strategies: (1) pursuing integration and coordination, (2) bargaining collectively, and/or (3) maintaining and improving the open market.

Integration and coordination are the cooperative version of current strategies used by most large-scale, noncooperative food firms. It is a strategy of vertical integration; that is, combining under one management two or more stages of the production-processing-servicing-marketing system. Often it is expanded through other strategies to aid in coordination and market development.

Bargaining collectively is the strategy of cooperative market power. It is a policy base for "self-help" types of programs for maintaining the family farm production unit protected by a strong power-base bargaining organization. This is group action to effect countervailing power, usually with public sanction and often built around a commodity.

Maintaining and improving the open market is the strategy of providing an environment that allows cooperatives to provide necessary services in a highly competitive system, usually at the first-handler level. Concepts underlying this strategy are traditional to cooperatives and depend heavily on the open competitive market with open information and a large number of buyers and sellers. Innovations for improving open markets include detailed information systems and exchange mechanisms such as the teleauction.

These three strategies point the way for an individual cooperative and guide policymakers for cooperatives in general. Unfortunately, these strategies do not provide a recipe for an individual cooperative. Individual cooperatives must adjust and adapt these general strategies to their own situations. To do this, they must understand and utilize the more specific strategies that make up these general strategies.

Organizational strategies used to accomplish the general strategies include (1) horizontal coordination, (2) diversification, (3) multi-cooperative organizations, and (4) joint ventures. These strategies are typical management devices by which an individual cooperative can adjust to the environment.

Facilitating strategies needed include (1) acquiring or maintaining government sanction (such as the Capper-Volstead Act), (2) using information systems, (3) enhancing domestic demand, (4) expanding foreign trade, and (5) expanding financial sources. These strategies are sometimes not under the control of the individual cooperative and may require group action. Such action may be taken jointly by several cooperatives or may be instigated by the public at large. Most of these arrangements are external to the firm.

Planning

Planning is the hard work necessary to see a strategy come to life. Without planning, a strategy is useless. Planning requires imagination, time, and a commitment.

Agribusiness planning has been dominated historically by a strong leader. Usually this was because (1) he owned much of the business, (2) he had developed the patent(s) or brands that provided the lifeblood of the business, or (3) he had founded the business. Likewise, most individual cooperatives have had one or a few dominant personalities. These people held this role because of their overall understanding of the cooperative's operations, their historical perspective of the cooperative, their producer loyalty, and/or their knowledge of the supply sources or the product outlets of the cooperative. The size and simplicity of the cooperative made it possible for one or a few people to plan fairly adequately. Cooperative success attests to this. But external and internal complexities of the future will require more planning.

Formal planning is now done by most large noncooperative businesses. Despite the fact that many cooperatives think they have adequate planning, the authors found that they lag behind noncooperatives in the total corporate planning process. However, a few large regional cooperatives are becoming increasingly involved in total corporate planning.

A long-range plan should include three types of plans—operational, project, and strategic—all coordinated and combined.

Operational plans regard the future of existing activities in existing markets with existing customers and facilities. Most large cooperatives studied did this type of planning; some called it a budget and others a plan. Usually such planning was not well done.

Project plans include specific projects that are expected to take the cooperative into something new. Cooperatives did this type of planning for large projects, but smaller projects often were given little planning.

Strategic planning is fairly new and is a broad concept. It specifies the kind of cooperative the board and members want, the purpose and scope of the cooperative, the cooperative goals, the strategic alternatives for attaining these goals, and the operational details to carry out the strategy. These items are put in writing and quantitative data are used when possible. Top management and key board officials must be deeply involved.

The Board and Management

Board-management relationships hold many of the keys to the survival of cooperatives. The creativity rests here; much of the knowledge needed should already be ingrained in the minds of these people; the strategic sense of good business management must be inherent; and leadership must be dedicated and committed. Without these, no strategy or plan can assure cooperative survival.

This book did not evaluate the board-management relationship in depth, but it focused on the cooperatives that were recognized to have qualified management-board teams. In this light, many of the things being done by these cooperatives indicate the type of board-management relationship that is necessary.

Cooperative management continually must be upgraded. Evidence of substantial upgrading was found in several of the progressive cooperatives. Others should follow this lead. Salaries of cooperative managers have increased relative to those in noncooperatives, but cooperative salaries still lag. Several programs for improved management training and compensation are being tried by cooperatives. Depth of management was of concern to several managers.

Board involvement varied substantially among the cooperatives. An ESCS, USDA, study on directors provides additional guidelines for board evaluation and improvement. The study—a detailed survey of 1,833 cooperative directors throughout the nation, provides profiles of their background; of their work as directors; and of some of their thoughts, attitudes, and perceptions.[1]

The Knowledge Needed

One purpose of this book was partially to meet a need for a more integrated, up-to-date compilation of appropriate knowledge needed by cooperative leaders building strategic plans. Information in this book is broad and detailed for that purpose.

The breadth of knowledge needed is wide and continually changing. Eight major components and their respective elements expected to shape the direction of cooperatives were identified (Table 8.1). The most likely outcome is projected for the next ten years.

TABLE 8.1. The major components and elements affecting farmer cooperatives and the most likely outcome, 1978–87

Major components and elements	Most likely outcome
Structure	
Farms:	
Number	declining
Control	family*
Size	increasing
Capital requirements	increasing
Farmers' attitude toward cooperation	same
Member commitment	increasing†
Cooperatives:	
Locals	fewer in number and larger in size
Centralized regionals	same in number and larger in size
Federated regionals	fewer in number and larger in size
Mixed federated and centralized	increase in number and larger in size
Commercial interregionals	increase in number and larger in size
Noncommercial cooperative organizations†	same in number with increased responsibilities
Joint ventures (co-op with non-co-op)	slight increase in number
Cooperative Functions	
Inputs:	
Merchandising only§	decrease in importance
Manufacturing	increase in share of farm needs
Raw material	increase in share of farm needs
Marketing:	
First handling only	increase in market share
Processing	increase in market share
Sale to food service institutions	slight increase in market share
Food retailing	same in market share
Bargaining:	
Raw farm products	same in market share
Services	increase in share or use of new services

TABLE 8.1. *(Continued)*

Major components and elements	Most likely outcome
Competition	
Handling of farm supplies:	
Competition from other cooperatives	slight increase
Competition from noncooperatives	slight decrease
Marketing of farm products:	
Competition from other cooperatives	slight increase
Competition from noncooperatives	increase
Geographic Marketing Orientation	
Cooperatives:	
Locally	same
Regionally	increase
Nationally	slight increase
Internationally	slight increase
Level of Management	
Board policy	increase in progressiveness
Management compensation	more competitive
Management depth	slight increase
Research and development	slight increase
Planning‖	increase
Capital	
Member equity	decrease in total share of financial sources
Banks for Cooperatives	slight decrease in share of total debt
Commercial banks	slight increase in share of total debt
Issuance of debt securities	same share of total debt
Other#	slight increase
Public Policy	
Individual cooperatives	slight increase in policy involvement
Noncommercial cooperative organizations**	increase in policy involvement
Individual cooperative lobbyists	increase in use
Cooperatives' joint effort with:	
Commodity organizations	slight increase
General farm organizations	slight decrease
Farm price and income policy:	
Free market (export-dependent)	slight decrease in emphasis
Supply control and price support programs	slight increase in emphasis
Specific commodity programs	increase in emphasis
Public Sanctions	
Capper-Volstead Act	amendment and/or stricter interpretation of original act
Bargaining	continue with current legislation
Marketing orders	continue with current legislation
Marketing boards	no interest in enabling legislation
Tax policy	continue with current legislation

* Over the next decade, this appears most likely. The second decade may see important increases in corporate control.

† Member commitment will increase, mainly because of increased use of marketing agreements and contracts.

‡ Includes such groups as cooperative trade or research associations, e.g., the National Council for Farmer Cooperatives or Cooperative Research Farms.

§ Buying manufactured inputs for sale to patrons. Does not include the manufacturing of inputs.

‖ Formal long-range corporate planning.

Includes loans from insurance companies, leveraged leasing, leasing, and other long-term financing methods.

** Includes organizations such as the National Council for Farmer Cooperatives.

Individual cooperatives that are committed to planning will need knowledge of these components and elements. Many interrelationships among elements must be taken into consideration. For example, a declining number of farms may require fewer but larger local cooperatives, may result in increased competition from noncooperatives and other cooperatives, and may require greater public policy involvement by cooperatives.

The Future

Elements most crucial in determining the future of farmer cooperatives will include (1) farmers' attitudes toward cooperation, (2) member commitment, (3) noncooperative competition in the marketing of farm products, (4) the skill and progressiveness of boards and management, (5) the availability of capital, (6) the involvement of cooperatives in general public policy issues, and (7) the maintenance of cooperative public sanctions (Table 8.1).

STRUCTURE. Farmers are traditionally independent business executives, and their attitude toward cooperation is not expected to change significantly. Some increase in member commitment may occur as cooperatives develop more marketing pool programs. Growers will enter these programs because they will be considered an improvement over deteriorating open markets and an opportunity for participation in vertically integrated operations.

Cooperative structure is expected to continue its past trend—fewer numbers but increasing size. Merger and consolidation activity among local and smaller regionals will continue gradually to reduce cooperative numbers. The objective will be to reach more efficient operating size and to reduce overlapping of cooperative services. Horizontal merger activity among large regionals will probably be limited as a result of legal constraints. Some regionals will move further toward centralization, at least in some functions such as management of locals. However, the federated concept is likely to remain the most common structure.

Interregionals will be increasingly used for foreign trade, fertilizer production, energy search and production, joint ownership of transportation systems, and possibly the distribution of processed farm products. Such arrangements are not likely to be used widely for the sharing of market information (except for bargaining groups) or for joint sales of processed farm products. Joint ventures between cooperatives and noncooperatives may increase slightly (very few exist today). The many problems inherent in these relationships will limit their use. Cooperative leaders are likely to show more interest in joint ventures among themselves.

COOPERATIVE FUNCTIONS. Supply cooperatives are expected to continue to become more basic in the production and ownership of farm inputs. Increasing scarcity of resources and fluctuating prices will make this necessary. Increased use of interregionals will make it possible.

Cooperatives will experiment with more vertical integration in food processing. Four main reasons support this. They may need to (1) maintain a market outlet for some products; (2) enhance their share of the consumer's dollar; (3) maintain some countervailing strength; and (4) maintain consumer acceptability of farm-produced goods.

Because of the high capital requirements, vertical growth will be primarily restricted to large regional cooperatives. The vertical growth that does occur will occur through acquisition of noncooperative firms that are already in the market. This approach will be taken to reduce the high risk and substantial cost of market development.

Market share of agricultural product processing is likely to increase somewhat. But aggressive competition from noncooperative firms could easily reverse that trend. There will be increased interest in serving the food service industry because of its importance in insuring access to the consumer. Indications are that cooperatives will not vertically integrate into the food retailing level.

In both the supply and marketing fields cooperatives increasingly will diversify their operations but will continue their restriction to agricultural-related businesses. The only exceptions may occur in the marketing of farm type inputs to customers living in suburban areas. More and better coordination of supply and marketing functions is likely. Such coordination will be necessary to support new marketing activities and to assure markets for farm inputs. The full-service cooperatives, supplying one-stop service for farm inputs, marketing facilities, capital, and services will probably become more popular because of potential economies and because patrons will expect it.

The growth of bargaining cooperatives will be highly dependent on general farmer support for national and state bargaining legislation. Since grower support shows no significant signs of increasing, growth in this area is expected to be limited and may not add to the current share of farm products sold in this way.

COMPETITION. In the handling of farm supplies, cooperatives may experience a decreasing level of competition from noncooperatives.[2] This trend is expected for several reasons: the well-established delivery system that has been built up by supply cooperatives; the increased number of customers whom cooperatives were able to obtain during the fertilizer shortage; and the recent discontinuance of farm-related activities by noncooperative firms in response to more profitable opportunities outside agriculture.

Competition between cooperatives and noncooperatives in the marketing area will increase. This will occur because of the increased vertical integration and market share that cooperatives are likely to achieve. Increasingly, cooperative marketing firms will find themselves competing with the large conglomerates whose overall financial and other managerial strengths are likely to be superior. This will be especially true in new and expanding businesses such as the food service industry.

Some regional cooperatives will find slight increases in competition among themselves both in the supply area and in the marketing area. Most regionals will probably respect their traditional membership areas, but a few large ones will take the offensive in expanding into marketing areas they believe to be inadequately served. Competition for volume and a need to diversify may also be factors in increasing competition among cooperatives. Regulatory trends may be another factor. In the past, cooperatives finding themselves providing similar services in the same market occasionally decided to merge their operations. Such horizontal merger activity will be increasingly difficult, especially for large regionals. But mergers that combine completely separate functions will not be affected. Local cooperatives will probably continue to merge, with limited legal restraints.

GEOGRAPHICAL MARKET ORIENTATION. The emphasis in farm product marketing will be in local or regional markets. Severe competition and high capital requirements will severely limit most cooperatives from entering national markets, although some large aggressively managed regionals may increase their thrust in this area.

International markets will continue as important to United States farmers as in the recent past, and some efforts to enter them will be made by individual cooperatives. But the volume, capital, and managerial expertise necessary will encourage the use of multi-cooperative organizations for this purpose.

LEVEL OF MANAGEMENT. Cooperative boards and management will rapidly become better trained and more progressive as the competitive environment becomes tougher and cooperatives become larger. Operations will also be less on a traditional cooperative basis and more like noncooperative businesses. Salary differentials between cooperatives and noncooperative firms will narrow but will persist, except in the largest cooperatives. Some progressive farm cooperatives will find ways of tying management compensation to company net margins. Selecting top managers from within will allow few opportunities for experienced personnel from noncooperative firms to enter the upper levels of cooperative management except in the very largest organizations (often the interregional).

Cooperative directors and managers will gradually recognize the benefits of long-range planning, and its use will increase. Managerial capabilities will be upgraded in order to deal with the complexities of planning five to ten years ahead and to control business activity so that such plans are achieved.

CAPITAL. Both the large total capital needs of individual cooperatives and the strong competition for "on-farm" capital will serve to continue past trends toward an increased debt-to-equity ratio for cooperatives. As a result, the problems in the capital area may become directed more toward availability of equity capital than toward debt. To secure and retain needed equity capital, many cooperatives may be required to move away from traditional cooperative financing methods. Capital plans may include the use of more permanent member capital in place of revolving funds and will be designed so that the current member shoulders the equity capital burden. Activities in the marketing area that directly benefit a particular group of farmers increasingly will use "front-end" capital provided largely by the group affected.

The growth in Banks for Cooperatives system net worth is not expected to keep up with capital needs of very large cooperatives, although the use of advanced capital plans will provide ample lending capacity for most cooperatives. As a result the Banks for Cooperatives will continue gradually to lose in share of overall debt provided to cooperatives. Large cooperatives will move partially to alternative sources of capital to overcome the credit limits of the Banks for Cooperatives. But small cooperatives are not expected to reduce their reliance on this lending source.

Large cooperatives increasingly will be sought by investment banking firms as well as by commercial bankers. Treasurers will be forced to increase their level of knowledge and use of capital from alternative debt sources. Commercial banks may become a more important source of capital because of the additional services they can offer, especially in export financing. However, the number of financially related services offered by the Banks for Cooperatives is expected to increase, and this should reduce the attractiveness of commercial banks.

PUBLIC POLICY. Cooperative leaders will become more involved in trying to influence public opinions about cooperatives. The controversy over public cooperative sanctions and the increasing involvement of the government in business affairs (energy, transportation, and environment) will encourage more participation of cooperatives in forming and backing legislative efforts of the National Council of Farmer Cooperatives. Cooperatives will also be forced to boost their individual lobbying efforts.

Although individual cooperatives currently show little concern over farm policy legislation, cooperatives may gradually become more involved in designing or shaping these policies. The declining farm population will place a greater responsibility on cooperatives. The increased national and international sales activities of regional cooperatives will encourage greater input into foreign trade policies. This public policy involvement will be direct and through commodity groups. One reason that commodity groups may be used is the expected trend toward commodity-oriented farm policies. Joint efforts with the general farm organizations will decrease as cooperative management seeks to separate the social aspects of farm groups from the cooperative's business interests. Commodity-oriented cooperatives and general farm organizations should look for mutual interests where they exist, and should guard against unwarranted competition between the two groups.

PUBLIC SANCTIONS. Public sanctions are likely to come under tough scrutiny from both the public and noncooperative competitors as cooperatives increase in size, diversification, and area served. No new major sanctions are likely to be sought or gained for cooperatives. Existing sanctions will probably remain but will be interpreted more narrowly and cooperatives will be watched more closely than in the past. Managers were concerned that the Capper-Volstead Act might be brought up for study and that it might receive minor revisions by Congress. Areas affected would be monitoring for undue price enhancement and the "marketing agencies in common" provision. Tax administration under existing laws will continue a narrow interpretation as taxing authorities attempt to increase revenues through elimination of benefits bestowed on certain organizational forms. Current attempts to define what a cooperative is and how it should be operated are further indications of a future restrictive tax policy.

Expected Changes with Different General Strategies

The relative emphasis placed on many factors covered in this discussion of the future will be affected by the relative degree to which cooperatives accept and modify the three general strategies.

The *integration and coordination strategy* would put many stresses on cooperatives. The number of cooperatives would be greatly reduced, with large regional and interregional cooperatives dominating. Competition with large agribusiness conglomerates would be more direct. Competition among regional cooperatives would increase. Vertical integration would obviously increase, and pressure for growth would increase substantially. Producer loyalty and equity problems would grow and more use would be made of marketing agreements. The major issue would be how far to integrate into processing. Another key issue would be the level of emphasis on consumer entree. General farm organizations would be concerned by this commercial, seemingly nonfarm, emphasis. Markets would widen, and ability to handle foreign marketing would improve. Management sophistication would need to be increased substantially with much more thorough planning, especially of large, new ventures. Capital needs would increase, and sources would be more diversified and innovative. Public scrutiny would increase, and current public sanctions increasingly would be challenged.

The *bargaining collectively strategy* would change the orientation of cooperatives drastically. Where cooperatives have historically been reactors, they would need to become countervailing powers. Often this would mean moving before the competition does and with more market control. The decision as to who would be the bargaining agent would be a severe problem. Larger cooperatives, probably interregionals, would be needed to make national bargaining programs effective. Some special crops and products would be bargained by smaller, more specialized bargaining cooperatives, much as now. Cooperative functions would be added in processing and distribution to protect the bargaining power. Dairy cooperative growth is a good example. Growth was originally to enhance bargaining efforts and has since developed using the integration and coordination strategy. Competition would view such a strategy as a direct power confrontation and would resist it.

Bargaining cooperative management would be more specialized, with particular emphasis on techniques of negotiation, use of market information, and political involvement. Capital requirements would not be as great as for the integration and coordination strategy. If public policy programs shifted in the direction of bargaining, this would markedly increase emphasis on the bargaining strategy. But that is not expected unless the economic status for farmers deteriorates substantially. The self-serving nature of such public sanction would be greatly challenged by other groups, especially organized consumer groups.

The *maintaining and improving the open market strategy* has much philosophical appeal to farmers. It is consistent with their conservative political and economic history. It is the basic orientation of much early cooperative activity. But this emphasis would be in opposition to most changes in business structure and organization occurring in recent years. Modern management has reduced the number of buyers and sellers, reduced the number of price-making points, has put more marketing functions under control, and has probably concentrated market information into fewer hands. To reverse these trends would be a major task for cooperatives even as an organized group. Cooperatives have some special public sanctions and possibly some legal and other comparative advantages which would allow them to reverse the movement away from the open market concept. They are relatively free to work on the price-making area, the information-gathering area, and interfirm cooperation. The public might support these activities. Noncooperative businesses would resist them. Management levels would need to be increased over current levels in cooperatives following this strategy, but such a strategy would not require the management sophistication nor the capital requirements needed for the integration and coordination strategy. Any major movement toward an open market farm policy would favor the open market cooperative strategy, but the trend appears away from this.

Strategic Implications for the Individual Cooperative

The strategy must be appropriate. Each cooperative has "roots." It has a historical orientation. Its members founded it for a reason. They have stuck by it and it has survived for a set of reasons.

Every cooperative has a record of change. No surviving organization can stay fixed, unchanging. The cooperative has added facilities, services, and products. It got where it is by some definable processes.

Every cooperative has a geographic area of concentration. People know the area it serves. It has kept a close-knit family, or it has expanded to add widely different members over a broad area.

Every cooperative has a way of relating to other organizations. It is a loner or it tends to be a joiner. It is involved broadly in community activities or tends to its own business.

A strategy must be appropriate for the unique nature of a cooperative. It must move the organization from its distinct and current uniqueness to a new posture to which it can continue to build. The managers interviewed had a great interest in what others were doing and planning. Farmers and managers must study cooperative strategies used by others, but they must not simply mimic other strategies. Their strategy should synthesize ideas but retain the unique comparative advantages of their individual cooperatives.

The strategy must be decisive. The notion of strategy is to make a decision—a decision about what you want to achieve. Decisiveness must have a sharpness—something new. Precision, definiteness, direction, and quantification should be characteristics of such a strategic decision.

The historical purpose and resultant strategies of cooperatives may have been rather simple and straightforward. More care will be required to fine-tune these for the future.

Choice is necessary. Once a general broad strategy is selected, other strategies must come from choices among a range of organizational and facilitating strategies, alternative cooperative sizes, alternative markets, and alternative products.

Decisiveness gives understanding to members, board, management, and others who do business or relate with the cooperative.

A decisive strategy is a prerequisite for total planning. The rigor of precision is a growing component of modern planning. For example, a computerized planning model is a precise instrument and will not tolerate indecisive, sloppy, poorly defined strategies.

The strategy must be balanced. Probably the techniques for balancing alternatives are the most effective aspects of modern planning processes. Computer and quantitative techniques have added much here. For example, equity considerations of patrons are a most significant issue of balance. What are the effects of the strategy on services rendered to the small patrons versus those rendered to the larger patrons? Where should the equity capital be raised, from whom, for what?

The strategy has to be balanced in expectation about other cooperatives or noncooperative competitors. Is there enough business to go around and to make the strategy work? Does the strategy place too much risk on the membership?

A balanced strategy is required for effective use of resources. This may be difficult. Does the strategy effectively use the facilities available or readily accessible? Does it utilize existing or planned management capability?

The strategy must be consistent. One of the main causes of business failure is vacillation. The strategy must have staying power over time. The cost of projects and decisions on the drawing boards of the cooperatives studied preclude "fits and starts." Past mistakes and redirections have been relatively inexpensive compared to the possible expense of mistakes in the future.

The consistency must have an aspect of adaptability. Changes will come. Options can be set up which are consistent with the total strategy.

When conditions change, a new but consistent option should already be designed and ready to be implemented without cause for a whole new strategy.

Control techniques require consistency; otherwise no measure of performance is meaningful. Planning and control assume change in plans. Comparisons of results are to be made against plans. Controls should show performance and particularly problems caused by change. This is the essence of consistency in planning.

Farmers do not change rapidly; consistency with properly planned direction appeals to them.

Strategic Implications for Cooperatives in General

The strategy must be assertive. Many things argue against a status quo approach by farmers and their cooperatives. Farmers must have their cooperatives do something. Several reasons explain this—for example, the projected environment, expectations of farmers, public pressures on farmers and their cooperatives, basic survival issues of farmers and their cooperatives, and potential gains for farmers.

Many cooperatives are experimenting with new assertive strategies. Some individual experiments obviously have important consequences for the whole cooperative movement—some deserve support, some do not. Possibly the time has come for cooperative leaders to keep each other more informed about the range of experimentation going on so the lessons learned by one cooperative can be quickly known to others.

The strategy must be timely. The temptation always exists to say the current time is crucial. But this book concludes that the time is ''propitious'' and some element of urgency does exist.

Several people interviewed, including cooperative managers, thought that the decade of the 1980s in many ways will be the decade of choice for cooperatives. Yet a positive response was not evident. For example, farm structure analysts agree that farm structure can be radically changed within a decade. One proposed definite program to effect this is a cooperatively controlled agriculture. Yet this study found few people concerned with this point.

Timeliness of a strategy is important for several reasons. For example, if farmers and their cooperatives are to survive, they need assured entree to the final consumer markets, assured access to scarce sources of raw materials, maintenance of the demand for farm-produced food ingredients, continued retention of appropriate public sanctions, and a shared portion of expanding foreign markets.

The strategy must be oriented to other cooperatives. Managers did not display great concern for issues they had in common with other cooperatives. Preoccupation with individual immediate problems is natural. Yet enough concern was shown to suggest that cooperative strategies increasingly will recognize the position of other cooperatives.

Encroachment of membership areas by other cooperatives concerned some. Some thought the experiments in multi-cooperative efforts have proved themselves. Others had doubts. Some believed that cooperatives must pick up new responsibilities of public policy and public relations for agriculture.

Most managers were thinking primarily about their cooperative and where it was going. All these actions add up to a strategy that may affect other cooperatives. To create this effect by default or happenstance is a poor strategy. Strategies should be oriented to other cooperatives. Marketing cooperatives have a comparative legal advantage in this area.

The strategy must be producer-oriented. Managers and others involved in this study were most emphatic in saying that the most important aspect of cooperative strategy was the owner-patron relationship. Managers often reaffirmed that no strategy could violate the basic and first responsibility of cooperatives to their owner-producer members. This was true whether the issue was changes in product line, level of integration, joint ventures, or management capability.

The public historically justified cooperatives on the grounds of producer needs to cooperate. The public environment, predicted in this book, will closely monitor that special right. A few violations of this concept by individual cooperatives could have important consequences for cooperatives in general.

The strategy must be socially responsible. The current environment causes important concerns about the continuation of public sanctions for cooperatives. But this book concludes that responsible cooperatives will have little fear of losing much of their unique public treatment in the next decade. It also concludes that the policymakers in the projected social and political environment will not tolerate cooperative strategies that are socially irresponsible.

The public will monitor cooperative characteristics such as farmer orientation, reasonableness in size and scale, priority on orderly marketing as contrasted to price enhancement, joint activities with nonfarm oriented businesses, and maintenance of a broad-based membership structure. Farmers must take the responsibilities to see that cooperatives in general establish socially acceptable strategies.

Good cooperative leadership may not be enough to assure survival. Cooperatives may well need leaders with statesmanship. The destiny of cooperatives will be in their hands, and these must be capable hands. Such leadership will have great demands from many places in agriculture, and these demands will compete with cooperatives for the time of top agricultural people. Competition will be stronger. Cooperatives increasingly will compete among themselves. Growth at times will be difficult to achieve. Special privileges will be limited. More problems will require joint approaches. All these developments put pressure on leadership. The complexities, subtleties, and human problems of such developments require sophisticated knowledgeable leadership. Yes, even statesmanship if cooperatives are to survive.

APPENDIX TABLES

TABLE A.1. Cooperatives and other organizations represented and individuals interviewed for book, 1975–76

Names of organization and individual	Title or position
American Institute of Cooperation:	
Owen K. Hallberg	President
Agricultural Council of California:	
Leland Ruth	Executive Vice-President
Donald G. Gordon, Jr.	Governmental Relations Specialist
Agway Inc.:	
Ronald N. Goddard	Executive Vice-President and General Manager
Glen E. Edick	Group Vice-President, Staff Services
Alexander J. Kish	Vice-President, Treasurer
Carlton C. Dennis	Vice-President, Planning
Paul Taber	Vice-President, Public Relations
Adrian P. Driggs	Vice-President, General Counsel
Darwin G. Braund	Dairy and Livestock Research Director
Berkeley Bank for Cooperatives:	
William Pendred	President
Earl Brown	Vice-President
Blue Anchor, Inc.:	
Walter M. Tindell	President
Calcot:	
G. L. Seitz	President
California Canners and Growers:	
Robert L. Gibson	President
W. James Washburn	Vice-President–Assistant to President
Henry Schacdt	Secretary
Richard C. Cunan	Assistant Secretary
California Canning Peach Association:	
Ronald A. Schuler	President
California Tomato Growers Association, Inc.:	
Robert F. Holt	Executive Vice-President
Evan Hushbeck	Assistant Manager
Central Livestock Association:	
Kieffer Lehman	President and General Manager
Francis Anderson	Executive Vice-President and General Manager all C.O.B. Operations
Donald Staples	Hog Manager and Head Salesman
A. L. Sayers	Communications Director

TABLE A.1. *(Continued)*

Names of organization and individual	Title or position
CF Industries, Inc.:	
R. R. Baxter	President and Chief Executive
Donald V. Borst	Executive Vice-President
Donald R. Clark	Vice-President, Corporate Planning and Development
Gerald J. Maynard	Treasurer
Citrus Central, Inc.:	
John St. John	Executive Vice-President
James R. Terrill	Vice-President, Administration & Finance
Cooperative League of the USA:	
Art Danforth	Secretary-Treasurer and Director of General Services
Diamond-Sunsweet, Inc.:	
William Libby	Production Manager
Farm Credit Administration:	
Gene Swackhamer	Deputy Governor
George Irwin	Director—Research Division
Don Osborn	(Missouri visiting professor)
Noel G. Stocker	(Supervisory group)
Frank D. Aigner	(Research Division)
Farmers Union Central Exchange, Inc. (CENEX):	
Virgil Knudson	Vice-President, Public Relations
Tom Quimbly	Assistant General Counsel
Marvin Wieland	Vice-President, Farm and Home Supply
Maurice Miller	Director, Corp. Budget
Harlan Rosvold	Vice-President, Engineering & Construction
Darrell M. Moseson	Senior Vice-President, Finance
Charles Habergarten	Vice-President and General Counsel
Ingvald Lee	Senior Vice-President, Operations
Jerry Hahn	Financial Service Manager
Noel Estenson	Vice-President, Financial—Services
Farmers Union Grain Terminal Association (GTA):	
Robert Handschin	(Economic Research)
Mel Werner	Vice-President, Grain Marketing
Harvey Kaner	Corporate Counsel
Robert Johansen	Director of Public Relations
Farmland Industries, Inc.:	
Ernest T. Lindsey	President
Gordon Leith	Corporate Vice-President
Don Ewing	Financial Vice-President
Buell W. Beadle	Vice-President, Research and Development
Bernard Sanders	Executive Director of Economic and Market Research
Bob Ferguson	(Finance)
FS Services, Inc.:	
E. V. Stevenson	Executive Vice-President and General Manager
Ward Lawler	Vice-President, Administration
Ross Gilfillian	Vice-President, Merchandising
Tom Mulligan	Vice-President, Finance
Howard Elder	Marketing Director, Corporate Planning
Gold Kist, Inc.:	
D. W. Brooks	Chairman of the Board
G. A. Burson	Executive Vice-President and Chairman
W. W. Gaaston	Senior Vice-President
John Moseley, Jr.	Secretary and Comptroller
John Eskew	Director of Laboratory Services
Joseph Marshall	Vice-President, Foods Group
J. E. Marion	Director, Corporate Research
Indiana Farm Bureau Cooperative Association, Inc. (IFBCA):	
Harold Jordan	General Manager
Glenn Franklin	Assistant General Manager
Prentice Cummings	Controller—Assistant Treasurer
Ivan Moore	Economic, Market Research Director

TABLE A.1. *(Continued)*

Names of organization and individual	Title or position
Chuck Mitchner	Assistant Controller
Robert Maxwell	Manager, Livestock and Poultry
John H. Stein	General Counsel
Jack Kelly	Research—Nutrition
Landmark, Inc.:	
Kenneth N. Probasco	Executive Vice-President and General Manager
Howard Runyeon	Vice-President, Controller
John Moore	Vice-President, Secretary-Treasurer
Kenneth Peterson	Vice-President, Grain
George Morrison	Vice-President, General Counsel
Raymond Casey	Assistant to Executive Vice-President
Land O'Lakes, Inc.:	
Ralph Hofstad	President
Harvey H. Ebert	Executive Vice-President, Food Processing and Marketing
Raymond Watt	Senior Vice-President, Agricultural Services
Rolf E. Haugen	Senior Vice-President, Finance
Richard H. Magnuson	Vice-President, General Counsel
Don Ault	Director of Corporate Planning
MFC Services (AAL):	
Jerry L. Harpole	President and General Manager
Robert A. Filgo	Senior Vice-President and Treasurer
James W. Bryan	Executive Director, Egg Operations
Troy Pitts, Jr.	Executive Director, Crop and Agronomy Operation
John Bard	Executive Director, Broiler Operations
Robert G. Ratcliff	Executive Director, Livestock Operations
L. D. "Pete" Stacks	Executive Director, Louisiana
William R. Scharman	Executive Director, Corporate Services
Milton Mitchell	Legal Counsel
National Council of Farmer Cooperatives:	
Kenneth D. Naden	President
Paul S. Weller	Vice-President, Affairs and Secretary
Donald E. Graham	Vice-President and General Counsel
Donald K. Hanes	Vice-President, Public Relations
Bill Brier	Director of Energy Resources
James S. Krzyminski	Assistant General Counsel (Transportation)
Ocean Spray Cranberries, Inc.:	
Harold Thorkilsen	President
Kenneth J. Beeby	Counsel
John S. Connolly, Jr.	Senior Vice-President, Marketing
George C. P. Olsson	Vice-President, Government Relations
Ontario Pork Producers Marketing Board:	
Dayne Peer	General Manager
Jim Boynton	General Secretary
Jim Rollings	Sales Manager
Pulaski Livestock Market:	
Roy Meek	Manager
Ed Eller	Extension
Kenneth Neal	Extension
Riceland Foods:	
L. C. Carter	President
Wilfred Carle	Executive Vice-President and General Manager
George Vickers	(Assistant to Wilfred Carle)
Charlie Gunnell	Assistant Controller
Jim Pollard	(Domestic rice sales and all consumer products, including shortening)
Dorothy Seidenschwarz	General Sales Manager, Soybean Meal
Sun-Maid Raisin Growers (Sun-Maid):	
Frank Light	President
Lee W. Halverstadt	Secretary-Treasurer

250

TABLE A.1. *(Continued)*

Names of organization and individual	Title or position
Sunkist Growers, Inc.:	
Roy Utke	President
Russ Hanlin	Vice-President, Products Group
Dave Hitchin	Vice-President, Administration
William K. Quarles, Jr.	Vice-President, Government Affairs
John M. VanHorn	Vice-President, Field Services
Paul Wilson	Treasurer
Lynn Jones	Corporate Information Officer
Linda Shepler	(Assistant to Lynn Jones)
Tri-Valley Growers:	
Al Spina	Senior Vice-President, Administration
Donald Schulak	Senior Vice-President, Finance
Western Farmers Association:	
O. R. Wiebe	President
R. A. Baldwin	Vice-President and Treasurer
John Lackland	Vice-President and General Counsel
Fred Wright	Vice-President, Member Services
Don Dewitt	Vice-President, Marketing
Ralph M. Folsom	Director of Internal Audit and Special Projects
R. C. Bell	Editor, Western Farmer Magazine
William O. Conn	Controller
Julie Weston	(Legal)
Individuals:	
Ralph Bunje	Agricultural Marketing Consultant
Eric Thor	Director of Communications, University of California
C. William Swank	Executive Vice-President, Ohio Farm Bureau Federation, Inc.
Clarence Paar	Director, on State Board—Potato Growers of Idaho, Inc.

TABLE A.2. **Cooperatives in selected core sample**

Name of cooperative	Headquarters
Agway, Inc.	Syracuse, New York
Farmland Industries, Inc.	Kansas City, Missouri
FS Services, Inc.	Bloomington, Illinois
Indiana Farm Bureau Cooperative Association, Inc.	Indianapolis, Indiana
Land O'Lakes, Inc.	Minneapolis, Minnesota
MFC Services (AAL)	Jackson, Mississippi
Riceland Foods	Stuttgart, Arkansas
Western Farmers Association	Seattle, Washington
Citrus Central, Inc.*	Orlando, Florida
Sunkist Growers, Inc.*	Van Nuys, California
California Canners and Growers*	San Francisco, California

* Answered certain questions related to marketing.

TABLE A.3. **Academic personnel in selected sample**

Name	University
C. H. Ingraham	Ohio State
E. Fred Koller	Minnesota
Emerson M. Babb	Purdue
Julian A. Raburn	Georgia
Leon Garoyan	Davis
Olan D. Forker	Cornell
Richard Vilstrup	Wisconsin
V. James Rhodes	Missouri
William E. Black	Texas A&M

TABLE A.4. Cooperative production of formula feed, by region, U.S., 1969 and 1975

| | Establishments | | | | Quantity | | Percent of Co-op | | Percent of total by all co-ops in region | |
| | Total | | Percent of total in region | | | | | | | |
Region	1969	1975	1969	1975	1969	1975	1969	1975	1969	1975
	number		*percent*		*tons*			*percent*		
Northeast	181	177	10.0	10.9	3,231,632	2,871,510	15.6	14.0	32.1	24.2
Lakes States	373	318	20.6	19.6	3,776,445	4,660,892	18.2	22.8	39.3	44.9
Corn Belt	663	538	36.7	33.1	6,146,089	4,750,422	29.6	23.2	28.0	24.6
Northern Plains	309	323	17.1	19.9	2,060,521	2,203,805	9.9	10.8	21.5	23.0
Appalachian	92	86	5.1	5.3	1,080,483	1,034,038	5.2	5.0	13.2	13.7
Southeast	26	43	1.4	2.7	1,205,537	1,454,397	5.8	7.1	13.1	15.3
Delta States	24	27	1.3	1.7	705,290	1,005,747	3.4	4.9	10.7	16.4
Southern Plains	56	33	3.1	2.0	657,796	513,568	3.2	2.5	6.1	4.1
Mountain	40	49	2.2	3.0	462,144	541,588	2.2	2.6	6.6	6.8
Pacific	46	29	2.5	1.8	1,435,332	1,450,175	6.9	7.1	16.4	15.0
Total U.S.	1,810	1,623	100.0	100.0	20,761,269	20,486,142	100.0	100.0	20.6*	19.6

Source: J. Warren Mather and John M. Bailey, *Cooperatives' Position in Feed Manufacturing*, FCS Res. Rpt. 25, FCS, USDA, Washington, D.C., Nov. 1973, and Carl J. Vosloh, Jr., *Structure of the Feed Manufacturing Industry, 1975: A Statistical Summary*, Statistical Bulletin No. 596, ESCS, USDA, Washington, D.C., Feb. 1978.

* Weighted average of all regions.

TABLE A.5. Grocery store sales by fifty largest U.S. nonpublic and publicly held grocery chains, ranked by size of sales, 1976

Rank	Retail firm	Sales	Rank	Retail firm	Sales
		$1,000			$1,000
1	Safeway	10,442,531	26	Cullum	524,998
2	A & P	7,235,854	27	J. Weingarten	457,239
3	Kroger	6,182,991	28	Arden-Mayfair	450,140
4	Lucky	3,525,036	29	Chatham Super Markets, Inc.	440,000
5	American	3,464,655	30	Hy-Vee Food Stores, Inc.	440,000
6	Winn-Dixie	3,265,916	31	H.E. Butt Grocery Co.	440,000
7	Jewel	2,981,429	32	Weis	402,917
8	Food Fair	2,507,040	33	Kohl's Food Stores	370,000
9	Grand Union	1,622,633	34	Furr's Inc.	358,000
10	Super Markets General	1,612,692	35	Thriftimart	338,498
11	Albertson's	1,490,839	36	Smith's Management Corp.	325,200
12	Stop & Shop	1,474,872	37	Foodarama	325,168
13	Fisher Foods	1,442,546	38	Pick-N-Pay Supermarkets, Inc.	316,000
14	National Tea	1,397,209	39	Niagara Frontier Services	295,068
15	Publix Super Markets	1,247,772	40	Marsh	294,800
16	Dillon	1,148,399	41	Alterman Foods	288,979
17	First National	992,635	42	Giant Eagle Markets	285,000
18	Colonial	979,211	43	Shop Rite Foods	283,235
19	Allied Supermarkets	901,625	44	Schnuck Markets	270,000
20	Giant Food	895,334	45	Shopwell	260,930
21	Fred Meyer	749,599	46	Tradewell Stores, Inc.	250,000
22	Waldbaum	744,793	47	Schwaegmann Bros. Giant Super Markets	250,000
23	Ralph's Grocery Co.	743,000	48	Supermarkets Interstate, Inc.	250,000
24	Borman's	622,221	49	Piggly Wiggly Southern, Inc.	230,000
25	Pueblo International	546,124	50	Seaway Foodtown	227,113
				TOTAL SALES	66,586,241

Source: "45th Annual Report of the Grocery Industry," *Progressive Grocer*, Vol. 57, No. 4, April 1978, pp. 112 and 116.

TABLE A.6. Summary of original general purposes of core sample cooperatives as taken from their respective Articles of Association

Summary of purpose	A*	B	C	D	E	F†	G	H	I	J
Provide specific activities mentioned	x		x	x	x	x	x	x	x	
Operate on a cooperative basis or plan	x			x					x	x
Provide economical methods for purchasing & distributing		x	x		x		x			
Cooperatively purchase & distribute		x			x					
Operate on a nonprofit basis	x					x				
Promote cooperation		x								x
Promote agricultural welfare & development		x					x			
Do what is necessary & proper for benefits of the association								x		
Promote general welfare of members					x					
Promote, foster, & encourage the business of producing farm products	x									
Do business for the mutual help & benefit of its shareholders, employees, & patrons										x
Provide orderly marketing							x			

* 1945 Articles of Incorporation.
† 1935 Articles of Incorporation.

TABLE A.7. Summary of current general purposes of core sample cooperatives

Summary of purpose	A	B	C	D	E	F	G	H	I
Improve farmers' economic position	x	x	x	x	x	x	x	x	x
Provide service, processing, & aggressive orderly marketing			x				x		x
Consider employees' welfare			x	x	x				
Ownership & control by members				x	x				
Well-being of the agricultural sector							x	x	
Coordinate activities to better serve members	x			x					
Provide quality inputs				x			x		
Promote welfare of the community				x	x				
Consider the welfare of general public			x	x					
Operate under sound business ethics & principles				x	x				
Promote general welfare of those engaged in agriculture						x			
Operate on a cooperative basis						x			
Provide quality products						x			
Maintain sound financial policies & practices					x				
Lend support to other farm organizations					x				
Meet and serve needs of agriculture			x						
Consider consumer preferences	x								
Engage in marketing & selling of agricultural products						x			
Supply products & services for farm production								x	
Market selected farm products								x	
Explore new financing & market endeavors where feasible				x					
Develop long- & short-term plans				x					
Provide science of management to member companies				x					
Expand sound & profitable agricultural enterprises	x								

TABLE A.8.   Differences between cooperatives and noncooperative corporations

Differences	Standard corporation*	Cooperatives
Purpose	To earn profits for investors; increase value of shares; and provide employment for owners of small corporations.	To maximize net and real income of member users; and provide goods and/or services at cost to member users.
	To serve the public generally.	To serve its members primarily.
Organization	Incorporated under state general corporation law; no federal charter.	Organized under state cooperative law; some such as federal credit unions under federal charters also.
	Except for closely held corporations, anyone may own stock.	Organized around mutual interests of its member users.
	Stock of large corporations is sold on stock exchange or over the counter."	No public sale of common voting stock—none listed on stock exchanges.
Control	By investors, the stockholders.	By member patrons.
	Policies determined by stockholders and directors.	Policies made by member users and directors.
	Voting on basis of stock ownership according to the number of shares held.	Voting in local associations usually on a one-man one-vote basis, or patronage basis. In federations, locals vote either on number of members represented or on volume of business done with the central organization.
	Proxy voting permitted. Frequently control is exercised by (families or) "inside cliques."	Generally, no proxy voting. Seldom that internal cliques can get control.
Sources of capital	From investing public.	From member users primarily.
	From successful business operations with all or part of the profits reinvested.	From net margins on succesful operations with reinvestment of part or all of them.
Distribution of net margins	To stockholders in proportion to number of shares of stock held.	To patrons on a patronage basis after modest dividends on stock have been paid; reserves and, in some cases, an educational fund and bonuses to employees are set aside.
Stock dividends	No limit—depends on amount of profits and distribution policy.	Limited to a nominal amount—generally does not exceed 8 percent.
Operating practices	Use conventional methods of financing—sale of stock, issuance of bonds, bank loans, and reinvestment of part or all of the profits.	Use revolving capital plan of financing based on the amount of business transacted with patrons in addition to conventional financing procedures.
	Usually purchase products on a cash basis.	Usually pool sales receipts and pay average prices by grade for products received.
	Business done with public generally and not restricted as to clientele except in exceptional cases.	Business done primarily—and in many marketing associations using marketing contracts exclusively—with members.
	Primarily interested in operational efficiency to cut costs; less interested in pricing efficiency.	Not only interested in operational efficiency but in pricing efficiency as well so that differential pricing by grades may reveal to producers ultimate consumer preferences, tastes, and purchases.
	Charge competitive prices or what "the traffic will bear."	Charge either competitive or "break-even" prices in purchasing associations.

TABLE A.8. *(Continued)*

Differences	Standard corporation*	Cooperatives
Initial transaction	The purchase or sale is a complete transaction.	The purchase or sale is, in a sense, a conditional transaction subject to a refund or additional payment at the end of the accounting period, if net savings are made.
Tax treatment	Subject to many kinds of taxes including state and federal corporate income taxes.	Also subject to many different kinds of taxes. However, cooperatives organized under Chapter 185 do not have to pay a state corporate income tax and, if about ten restrictions are met, can also be exempt from paying federal corporate income taxes. Net margins are taxable to farmer recipients.
	Privately owned electric utilities are subject to property taxes on real estate.	Rural electric cooperatives in Wisconsin are taxed on gross income rather than on assessed property values.

Source: Marvin A. Schaars, *Cooperatives: Principles and Practices,* A-1457, Cooperative Extension Programs, University of Wisconsin, Madison, 1971, P. 11.

* A corporate type of business.

NOTES

CHAPTER 1

1. Throughout the book, mention of cooperatives will refer to agricultural cooperatives unless stated otherwise.
2. J. Phil Campbell, "The Role of Research in Agriculture," Paper presented at the joint meeting of the Washington Academy of Sciences and the Helminthological Society of Washington, Washington, D.C., Nov. 20, 1975.
3. Martin A. Abrahamsen, "Research Practices and Problems of Farmers' Regional Associations," FCS Misc. Report 96, FCS, USDA, Washington, D.C., Feb. 1946.
4. Martin A. Abrahamsen, "Cooperative Research, Progress, Problems," FCS Research Report 26, FCS, USDA, Washington, D.C., Oct. 1973.
5. Cornell University, "Potential Increases in Food Supply Through Research in Agriculture," Special series of studies in grant report to National Science Foundation, 1976; Massachusetts Institute of Technology, Department of Nutrition and Food Science, *Protein Resources and Technology: Status and Research Needs,* Executive Summary, Washington, D.C.: National Science Foundation, Research Applied to National Needs, 1974; National Academy of Sciences, Board on Agriculture and Renewable Resources, *Enhancement of Food Production for the United States,* Washington, D.C.: NAS, BARR, 1975; National Academy of Sciences, Committee on Agricultural Production Efficiency, *Agricultural Production Efficiency,* Washington, D.C.: NAS, 1975; and National Academy of Sciences, *World Food and Nutrition Study—The Potential Contributions of Research* and five volumes of contributing papers, Washington, D.C., NAS, 1977.
6. The studies referenced in Footnote 5 all tend to make such evaluations, but cooperative leaders should also read much more general materials such as "Science at the Bicentennial: A Report from the Research Community," National Science Board, NSF, 1976.
7. USDA, "Research to Meet U.S. and World Food Needs," Report of a Working Conference, Vols. 1-3, Kansas City, Mo., July 1975.
8. Travis W. Manning, "The Agricultural Potentials of Canada's Resources and Technology," *Canadian Journal of Agricultural Economics* 23 (1, Feb. 1975): 17-18.
9. Ross Whitehead, "What'll We Eat in 1999?" *Industry Week,* May 17, 1976, pp. 30-33.
10. M. Rupert Cutler, "The Unexplored Frontiers of Nutrition in the 1980's," NIH Conference *The Research Basis of Clinical Nutrition: A Projection for the 1980's,* June 20, 1978.
11. National Academy of Sciences, *Climate and Food,* Board on Agriculture and Renewable Resources, 1976.
12. Helpful references on this issue are G. Edward Schuh, "The New Macroeconomics of Agriculture," *American Journal of Agricultural Economics* (Dec. 1976): 802-11; Leroy Quance, and Gary C. Taylor, "Anticipating the Future," *Looking Forward,* Economic

258

Research Service, USDA, Washington, D.C., Sept. 1977; *Economic Report of the President,* Jan. 1978, pp. 195-206; Robert G. F. Spitzer (ed.), "Agricultural and Food Price and Income Policy: Alternative Directions for the United States and Implications for Research," Report of a Policy Research Conference, Washington, D.C., Jan. 15-16, 1976; "Food and Agriculture Issues for Planning," GAO, CED-77-61, Apr. 22, 1977; and "International Food Policy Issues: A Proceedings," Foreign Agriculture Economics Report 143, USDA, Washington, D.C., Jan. 1978.

13. Excellent coverage of many related issues is in *Marketing Alternatives for Agriculture: Is There a Better Way?* Committee on Agriculture and Forestry, U.S. Senate, 94th Congress, 2d Session, U.S. Government Printing Office, Washington, D.C., 1976. This report covers market mechanisms concerned with determining price and maintaining market access for individual farmers; market institutions involving group action including use of cooperative marketing, bargaining, marketing orders, and boards; and broad regulatory proposals affecting the working of much of the economy including industrial restructuring and fine tuning with administrative actions designed to improve the present agricultural marketing system.

14. V. James Rhodes, "Role of Marketing and Procurement Systems in the Control of Agriculture," *Southern Journal of Agricultural Economics* (Dec. 1973): 32.

15. Kenneth R. Farrell, "Outlook for Food Supplies and Prices," *1976 U.S. Agricultural Outlook,* Senate Committee on Agriculture and Forestry, Dec. 18, 1975, pp. 47-54.

16. Helpful papers on this issue include Carol Tucker Foreman, "Toward a U.S. Food Policy," *1978 Food and Agricultural Outlook,* Senate Committee on Agriculture and Forestry, Dec. 19, 1977, pp. 10-20; and "National Nutrition Issues," GAO, CED-78-7, Dec. 8, 1977.

17. *Your Food: A Food Policy Basebook,* Publication 5, National Public Policy Education Committee and Cooperative Extension Service, Ohio State University, Columbus, Nov. 1975, p. 4.

18. USDA, "Consumers: A Restless Constituency," Report by Young Executives Committee, Washington, D.C., May 1974, pp. 8-9.

19. For a discussion of this area, see Nathan M. Koffsky, "What Has Happened Since the World Food Conference." *1976 U.S. Agricultural Outlook,* Senate Committee on Agriculture and Forestry, Dec. 18, 1975, pp. 117-21.

20. "Commitment to Development Through Cooperatives," A progress report of the overseas development programs of the Cooperative League of the USA, 1974.

21. Alfred Reifmen, "The Changing World Economy of the 1970's," *1976 U.S. Agricultural Outlook,* Senate Committee on Agriculture and Forestry, Dec. 18, 1975, pp. 13-29.

22. Leon Garoyan and H. M. Harris, Jr., "Industrial Restructuring: A Policy for Industrial Competition," pp. 91-103, and Ronald D. Knutson, Dale C. Dahl, and Jack H. Armstrong, "Fine Tuning the Present System," pp. 82-90, in *Marketing Alternatives for Agriculture.*

23. Dale C. Dahl and Winston W. Grant, "Antitrust and Agriculture," *Minnesota Agricultural Economist,* No. 574, Dec. 1975.

24. Knutson, Dahl, and Armstrong, "Fine Tuning the Present System."

25. Garoyan and Harris, "Industrial Restructuring."

26. *Your Food,* p. 25.

27. Don Paarlberg, "The Farm Policy Agenda," Address at National Public Policy Education Conference, USDA 2621-75, Sept. 11, 1975.

28. Harold R. Breimyer, "A Reconciliation of National Goals for Food and Aspirations of Farmers and Rural People," Paper given at Perry Foundation-UMC Seminar on Agricultural Marketing and Policy, Columbia, Mo., Dec. 5, 1975.

29. For one of the most concise statements of these options, see "Extreme Options for Future Structure of Agriculture," *Tar Heel Economist,* North Carolina State University, Raleigh, Aug. 1973.

30. Knutson, Dahl, and Armstrong, "Fine Tuning the Present System," p. 83. Also special note should be taken of the specific recommendations of these authors for the Department of Agriculture, pp. 83-84.

31. James L. Gulley, *Beliefs and Values in American Farming,* ERS-558, Economic Research Service, USDA, Aug. 1974, pp. iv and 57.

32. Frederick E. Webster, Jr., *Marketing for Managers* (New York: Harper & Row, 1974), p. 298.

33. Frederick E. Webster, Jr., "Does Business Misunderstand Consumerism," *Harvard Business Review* 51 (5, Sept.–Oct. 1973): 89.
34. Breimyer, "A Reconciliation of National Goals."
35. Select Committee on Nutrition and Human Needs, U.S. Senate, "Eating in America: Dietary Goals for the United States" (Cambridge, Mass.: MIT Press, 1977).
36. USDA, "Consumers: A Restless Constituency," Report of the Young Executives Committee, USDA, May 1974, p. 13.
37. USDA, "Consumers."
38. A rather comprehensive coverage of this occurs in Velma Seat, Margy Woodburn, and William T. Vastine, "Will It Be Good and Good for You?" *Your Food: A Food Policy Basebook,* pp. 36–48.
39. This problem is typified by the range of regulations on the subject. Refer to Seat, Woodburn, and Vastine, "Will It Be Good and Good for You?"
40. Quoted from James G. Patton, "Between You and Hunger," A Report of the Midwest Conference on Food Policy, June 6–7, 1974.
41. Martin A. Abrahamsen, "American Cooperation: Pioneer to Modern," *Farmer Cooperatives in the United States,* FCS Bulletin 1, Washington, D.C., Revised 1965, p. 50.
42. Bob Bergland, "A Brighter Day for America's Farmers," Remarks before the Independent Bakers Association, Washington, D.C., June 7, 1978.
43. Many references cited in this book can educate one to the road signs for such changes. These should be read, but one should also read good periodicals. One of the best periodicals for keeping abreast of the consumer-oriented end of this relationship is the *National Food Review,* published by the USDA.
44. Many accounts of the recent economic history for agriculture exist. The logic of this section draws heavily on two: Economic Report to the President transmitted to Congress Jan. 1978, pp. 195–206; and *Agricultural Food Policy Review,* USDA, ERS AFPRI-1, Jan. 1977, pp. 2–3.
45. "Food and Agricultural Policy Options," Budget issue paper, Congressional Budget Office, Washington, D.C., Feb. 1977, pp. xiii–xiv.
46. This is a controversial area and measurement is not easy. For another view, see Howard W. Hjort, "Statement before Subcommittee on Domestic Marketing, Consumer Relations, and Nutrition," Committee on Agriculture, U.S. House of Representatives, Washington, D.C., July 25, 1978.
47. Schuh, "The New Macroeconomies of Agriculture," p. 802.
48. Several analyses support this outlook, but considerable reliance was placed on that done by agricultural economists at Purdue University reported in P. R. Robbins, "Long-Range Outlook for Indiana Agriculture to 1985," Mimeo report, Nov. 1977.
49. Much is being written on this subject. Two broad articles about the future from somewhat contrasting views are D. G. Johnson, "World Food Outlook Improves," and V. W. Ruttan, "Who Will Feed Tomorrow's Hungry?" *Agenda,* AID, Sept. 1978, pp. 1–11.
50. "Toward a National Food Policy," Conference sponsored by Academy for Contemporary Problems, Dec. 14–16, 1975, p. 3.
51. GIST, U.S. State Department periodic publication, Aug. 1978.
52. Economic Report of the President, 1978, p. 198.
53. Bob Bergland, Remarks at White House Conference on Balanced National Growth and Development, Washington, D.C., Jan. 30, 1978.
54. Harold F. Breimyer, "What Will Agriculture Be Like in the Decade Ahead?" Presentation at Illinois Cooperative Leadership Conference, Champaign, Ill., Feb. 27, 1976.
55. Don Paarlberg, "Agriculture 200 Years from Now" as paraphrased in *Eastern Milk Producer,* Feb. 1976, p. 8.
56. "The Capital Crisis," *Business Week,* Sept. 22, 1975.
57. Henry C. Wallich, "Is There a Capital Shortage?" Presentation at the International Monetary Conference, Amsterdam, Netherlands, June 11, 1975.
58. Philip T. Allen, "Agricultural Finance Outlook, 1976," *1976 U.S. Agricultural Outlook,* Senate Committee on Agriculture and Forestry, Dec. 18, 1975, pp. 81–87.
59. Leroy Quance and Gary C. Taylor, "Anticipating the Future," *Looking Forward: Research Issues Facing Agriculture and Rural America;* and Cecil W. Davison and

Milton H. Ericksen, "Alternative Economic Settings for Agriculture: 1977–81," *Agricultural Food Policy Review,* ERS AFPR-1, Jan. 1977, pp. 12–29.

60. A concise statement of the longer run view is by Wayne D. Rasmussen, "Perspectives on Agricultural Policy," *1977 U.S. Agricultural Outlook,* Committee on Agriculture and Forestry, U.S. Senate, 94th Congress, 2d Session, Dec. 10, 1976, pp. 151–59.

61. J. B. Penn adapts the notion more recently in his article "Agricultural and Food Policy Beyond the Agriculture and Consumer Act of 1973," *1977 U.S. Agricultural Outlook,* pp. 160–67. The Congressional Budget Study, Budget Issue Paper, *Food and Agriculture Policy Options,* Feb. 1977, also has an economic policy orientation, especially in Chapters 1 and 2.

62. A bibliography of references on international interdisciplinary approaches has been prepared: "World Food Problems: An Interdisciplinary View," Iowa State University Library series in Bibliography No. 5, June 1976.

CHAPTER 2

1. *Marketing Alternatives for Agriculture: Is There a Better Way?* Committee on Agriculture and Forestry, U.S. Senate, 94th Congress, 2d Session, U.S. Government Printing Office, Washington, D.C., 1976.

2. USDA, *Food and Fiber System: How It Works,* Agriculture Information Bulletin 383, ERS, USDA, Washington, D.C., Mar. 1975, p. 15.

3. Ibid., p. 8.

4. John M. Bailey, *Potential for Cooperative Distribution of Petroleum Products in the South,* FCS Information 91, USDA, Washington, D.C., July 1973, p. 3.

5. USDA, *Structure of Six Farm Input Industries,* ERS-357, ERS, USDA, Washington, D.C., Jan. 1968, p. 5.

6. Ibid., pp. 5–6.

7. Alex F. McCalla and Harold O. Carter, "Alternative Agricultural and Food Policy Directions for the U.S.: With Emphasis on a Market-Oriented Approach," Preliminary report prepared for Policy Research Workshop on Public Agriculture and Food, Price and Income Policy Research, Washington, D.C., Jan. 15–16, 1976, pp. 11–13.

8. Lyden O'Day, *Growth of Cooperatives in Seven Industries,* Cooperative Research Report 1, ESCS, USDA, Washington, D.C., July 1978, p. 45.

9. McCalla and Carter, "Alternative Agricultural and Food Policy Directions," p. 13.

10. Crop Reporting Board, *Commercial Fertilizers: Final Consumption for Year Ended June 30, 1976,* SRS, USDA, Washington, D.C., Apr. 1977, p. 3.

11. Ibid., p. 6.

12. USDA, *Structure of Six Farm Input Industries,* pp. 28–29.

13. Duane A. Paul, Richard L. Kilmer, Marilyn A. Altobello, and David N. Harrington, *The Changing U.S. Fertilizer Industry,* Agricultural Economics Report 378, ERS, USDA, Washington, D.C., Aug. 1977.

14. USDA, *Structure of Six Farm Input Industries,* 32.

15. Formula feed contains two or more ingredients that are processed or mixed according to specifications.

16. Carl J. Vosloh, Jr., *Structure of the Feed Manufacturing Industry, 1975: A Statistical Summary,* Statistical Bulletin 596, ESCS, USDA, Washington, D.C., Feb. 1978, pp. 42 and 31.

17. USDA, *Structure of Six Farm Input Industries,* pp. 60–61; and USDA, *Food and Fiber System.*

18. J. Warren Mather and John M. Bailey, *Cooperatives' Position in Feed Manufacturing,* FCS Research Report 25, FCS, USDA, Washington, D.C., Nov. 1973, p. 15; and Vosloh, *Structure of the Feed Manufacturing Industry,* p. 42.

19. USDA, *Structure of Six Farm Input Industries,* p. 48.

20. Martin A. Abrahamsen, *Cooperative Growth Trends, Comparisons, Strategy,* FCS Information 87, FCS, USDA, Washington, D.C., Mar. 1973, p. 79.

21. USDA, *Structure of Six Farm Input Industries,* p. 44.

22. Abrahamsen, *Cooperative Growth Trends, Comparisons, Strategy,* p. 78.

23. History and Statistics, unpublished data, FCS, USDA, Washington, D.C.
24. USDA, *Agricultural Statistics 1977*, U.S. Government Printing Office, Washington, D.C., 1977, p. 434.
25. *Your Food: A Food Policy Basebook*, Publication 5, National Public Policy Education Committee and Cooperative Extension Service, Ohio State University, Columbus, Nov. 1975, pp. 56–57.
26. Ibid.
27. U.S. Department of Commerce, *Statistical Abstract of the United States 1975*, p. 739; and *Business Week*, Oct. 27, 1973, p. 78.
28. History and Statistics, unpublished data, preliminary.
29. Lloyd C. Biser, *Cooperatives' Farm Machinery Operations*, FCS Information 86, FCS, USDA, Washington, D.C., July 1972, p. 2.
30. USDA, *Handbook of Agricultural Charts: 1976*, Agricultural Handbook 504, Washington, D.C., Oct. 1976, p. 17.
31. USDA, *Agricultural Statistics 1977*, p. 468, Preliminary.
32. USDA, *The Future Role of Cooperatives in the Red Meats Industry*, Marketing Research Report 1089, ESCS, Apr. 1978.
33. Calculated from data in Richard G. Heifner, James L. Driscoll, John W. Helmuth, Mack N. Leath, Floyd F. Niernberger, and Bruce H. Wright, *The U.S. Cash Grain Trade in 1974: Participants, Transactions, and Information Sources*, Agricultural Economics Report 386, ERS, USDA, Washington, D.C., Sept. 1977.
34. Ibid., p. 7.
35. *Your Food*, p. 55.
36. Ronald L. Mighell and William S. Hoffnagle, *Contract Production and Vertical Integration in Farming, 1960 and 1970*, ERS-479, ERS, USDA, Washington, D.C., Apr. 1972, p. 4.
37. Randall E. Torgerson, "Be Ready to Grasp Export Opportunities," *Farmer Cooperatives*, FCS, USDA, Washington, D.C., Feb. 1976, p. 5.
38. Ibid.
39. W. Smith Greig, "A Description of Structural Trends in the Food Processing Industry in the U.S." in *The U.S. Food Industry: Description of Structural Changes—Volume I*, Technical Bulletin 129, Colorado State University Experiment Station, Fort Collins, p. 42.
40. Ibid., p. 31.
41. U.S. Department of Commerce, *Statistical Abstract of the United States, 1977*, 98th ed., U.S. Government Printing Office, Washington, D.C., July 1977, pp. 800–805.
42. Ibid., p. 808.
43. Ibid., p, 809.
44. Greig, "A Description of Structural Trends," p. 33.
45. O'Day, *Growth of Cooperatives in Seven Industries*, p. 21.
46. Gilbert W. Biggs and J. Kenneth Samuels, *Cooperative Fruit and Vegetable Processors in the United States*, Service Report 123, FCS, USDA, Washington, D.C., May 1971, p. 20.
47. O'Day, *Growth of Cooperatives in Seven Industries*, p. 16.
48. USDA, *The Future Role of Cooperatives in the Red Meats Industry*, p. 16.
49. Abrahamsen, *Cooperative Growth Trends, Comparisons, Strategy*, p. 9.
50. Greig, "A Description of Structural Trends," p. 37.
51. Ibid., pp. 37 and 40.
52. For institutional descriptions, see National Commission on Food Marketing, *Organization and Competition in Food Retailing*, Technical Study 7, U.S. Government Printing Office, Washington, D.C., June 1966, pp. 4–8.
53. "A supermarket is defined as any grocery store, chain or independent, with an annual sales volume of $1 million or more. A retail grocery chain is defined as 11 or more retail stores operated under common ownership." Bruce W. Marion, Williard I. Mueller, Ronald W. Cotterill, Frederick E. Geithman, and John R. Schmelzer, *The Profit and Price Performance of Leading Food Chains, 1970–74, A* study prepared for the use of the Joint Economic Committee, Congress of the United States, U.S. Government Printing Office, Washington, D.C., Apr. 12, 1977, p. 9.
54. Ibid., pp. 126–32.
55. *Your Food*, p. 50.

56. Ibid., p. 51.
57. W. Smith Greig, *The Economics of Food Processing* (Westport, Conn: AVI Publishing Co., 1971), p. 351.
58. *Food from Farmer to Consumer,* Report of the National Commission on Food Marketing, U.S. Government Printing Office, Washington, D.C., June 1966, p. 95.

CHAPTER 3

1. Bruce L. Swanson and Jane H. Click, *Statistics of Farmer Cooperatives, 1972-73, 1973-74, and 1974-75,* FCS Research Report 39, FCS, USDA, Washington, D.C., Apr. 1977.
2. History and Statistics, Cooperative Management Division, ESCS, unpublished data, USDA, Washington, D.C.
3. Edwin G. Nourse, *The Legal Status of Agricultural Co-operation,* The Institute of Economics (New York: Macmillan Co., 1927), pp. 30-32.
4. Ibid. Rochdale principles included the following: (1) The association would be a stock company, (2) members only would hold shares, (3) membership would require ownership of one share or more, (4) each member would sign rules of the association, (5) minimum purchase by each member from the association would be required each year, (6) a limitation of capital stock held by individuals in the association would be enforced, (7) sales would be for cash only, (8) competitive pricing would be practiced, (9) limited interest on capital would be paid, and (10) allocation of earnings, after deduction of expenses and funds for operation and education, would be made to patrons according to their patronage. Pp. 32-36.
5. Ibid., pp. 39-40.
6. Ibid., p. 45.
7. Ibid., pp. 400-401.
8. Purpose or objective will be used here as a proxy for reasons for being. Actually, reasons for being is a broad general concept. Purpose is a more specific concept and includes what a business wants to do, such as, "to purchase, sell, manufacture and market fertilizer." Objectives define ways the organization is to accomplish its mission. Goals, a subset of the objectives, are more specific and quantifiable. For a discussion of objectives and goals, see Leon Garoyan and Paul O. Mohn, *The Board of Directors of Cooperatives,* Publication 4060, Division of Agriculture Sciences, University of California, July 1976, pp. 55-71.
9. All names of the organizations studied have changed since their establishment. One organization resulted from the merger of two regional cooperatives. In this case, the purposes of both organizations were reviewed.
10. Nourse, *Legal Status of Agricultural Co-operation,* p. 49.
11. Marvin A. Schaars, *Cooperatives: Principles and Practices,* A-1457, Cooperative Extension Programs, University of Wisconsin, Madison, 1971, p. 10.
12. The Capper-Volstead Act, 7 U.S.C. 291, 292; 42 Stat. 388 (1922) Public-No. 146-67 Congress.
13. Marvin A. Schaars, "An In-depth Probe for Meaning of Cooperatives," *Blueprint for Cooperative Growth: Action, Innovation, Coordination,* American Institute of Cooperation, Washington, D.C., 1972-73, p. 306.
14. Jim Hightower, "Do Cooperatives Have a Social Responsibility?" Statement prepared for the Graduate Institute of Cooperative Leadership, Columbia, Mo., July 1974, pp. 2 and 9, mimeograph.
15. Harold Jordan, Speech given at annual meeting of Indiana Farm Bureau, Cooperative Association, Indianapolis, 1968.
16. "The Billion-Dollar Farm Co-ops Nobody Knows," *Business Week,* Feb. 7, 1977, pp. 54-64.
17. Frederick E. Webster, Jr., *Marketing for Managers* (New York: Harper & Row, 1974), pp. 4-5.
18. *The American College Dictionary* (New York: Random House), pp. 535-36.
19. Leon Garoyan, "Blueprints for Farmer Cooperative Action—In Cooperative Structure and Growth," *Blueprint for Cooperative Growth,* p. 2.

20. Richard W. Schermerhorn, "Feasibility Analysis: An Internal Part of Growth," A. E. Series 102, Paper presented at the annual meeting of the Idaho Cooperative Council, Idaho Falls, Nov. 1971, p. 2.
21. Fred Koller, "What Are Issues Involved in Cooperative Growth?" *Blueprint for Cooperative Growth,* p. 147.
22. Leon Garoyan and Gail L. Cramer, *Cooperative Mergers: Their Objectives, Success, and Impact on Growth,* Station Bulletin 605, Agricultural Experiment Station, Oregon State University, FCS, USDA, Feb. 1969, p. 4.
23. W. Smith Greig, *The Economics of Food Processing* (Westport, Conn.: AVI Publishing Co., 1971), Chapter 5.
24. Joseph W. McGuire, *Factors Affecting the Growth of Manufacturing Firms,* Small Business Management Research Report, University of Washington, Mar. 1963, p. 18.
25. Ibid., p. 18.
26. Robert D. Buzzell, Bradley T. Gale, and Ralph G. M. Sultan, "Market Share: A Key to Profitability," *Harvard Business Review* 53 (Jan.–Feb. 1975): 97.
27. McGuire, *Factors Affecting Growth of Manufacturing Firms,* p. 18.
28. Farmers Union Central Exchange, Inc., *A Country Get Together,* Annual Report, 1977.
29. "The Hard Road of the Food Processors," *Business Week,* Mar. 8, 1976, p. 50.
30. Alfred D. Chandler, Jr., *Strategy and Structure: Chapters in the History of the American Industrial Enterprise* (The MIT Press, Cambridge, Mass., © 1962 by The Massachusetts Institute of Technology), p. 19.
31. USDA, History and Statistics, Cooperative Management Division, unpublished data, ESCS, USDA, Washington, D.C.
32. Martin A. Abrahamsen, *Cooperative Growth: Trends, Comparisons, Strategy,* FCS Information 87, FCS, USDA, Washington, D.C., Mar. 1973, p. 87.
33. Lyden O'Day, *Growth of Cooperatives in Seven Industries,* Cooperative Research Report 1, ESCS, USDA, Washington, D.C., July 1978.
34. Philip Kotler, *Marketing Management: Analysis, Planning, and Control,* 3rd ed., © 1976, p. 49. Reprinted by permission of Prentice-Hall, Englewood Cliffs, N.J.
35. Ibid., p. 50. Reprinted by permission of Prentice-Hall, Englewood Cliffs, N.J.
36. Garoyan and Cramer, *Cooperative Mergers,* pp. 16–17; and Willard F. Mueller, *The Role of Mergers in the Growth of Agricultural Cooperatives,* California Agricultural Experiment Station Bulletin 777, 1966.
37. Edith T. Penrose, *The Theory of the Growth of the Firm* (New York: John Wiley & Sons, 1959).
38. Gilbert W. Biggs, *Farmer Cooperative Directors: Characteristics, Attitudes,* FCS Research Report 44, ESCS, USDA, Feb. 1978, p. 18.
39. A more in-depth discussion is provided under Expanding Financial Sources.
40. Ellen Haas, "Remarks of Ellen Haas," Remarks presented to the National Bargaining Conference, Washington, D.C., Jan. 12, 1976.
41. A more in-depth discussion is provided under Acquiring Government Sanction.
42. For more detail on taxation issues, see Taxation under Facilitating Strategies.
43. There have been some notable exceptions to this rule.
44. The most recent indication of this is contained in O'Day, *Growth of Cooperatives in Seven Industries.*

CHAPTER 4

1. A stage is considered as any operating process capable of producing a saleable product or service.
2. Ralph Trefon, "Guides for Speculation about the Vertical Integration of Agriculture with Allied Industries," *Journal of Farm Economics 41* (Nov. 1959): 736.
3. Ronald L. Mighell and William S. Hoffnagle, *Contract Production and Vertical Integration in Farming,* 1960, and 1970, ERS-477, USDA, Apr. 1972.
4. USDA, History and Statistics, Cooperative Management Division, unpublished data, ESCS, USDA, Washington, D.C.
5. Martin A. Abrahamsen, *Cooperative Growth Trends, Comparisons, Strategy,* FCS Information 87, FCS, USDA, Washington, D.C., Mar. 1973, p. 9.

6. Bruce Swanson and Jane Click, *Statistics of Farmer Cooperatives, 1972-73, 1973-74, and 1974-75,* FCS Research Report 39, FCS, USDA, Washington, D.C., Apr. 1977, p. 11.

7. For an evaluation of the use of subsidiaries by cooperatives, see Charles A. Kraenzle and David Volkin, *Subsidiaries of Agricultural Cooperatives,* Cooperative Research Report 4, ESCS, USDA, Washington, D.C., Feb. 1979.

8. The possiblility of forming a joint venture with a noncooperative firm is discussed under Organizational Strategies—Joint Ventures.

9. Peter Helmberger, "Future Roles for Agricultural Cooperatives," *Journal of Farm Economics 48* (5, Dec. 1966): 1431.

10. Theodore Levitt, "Marketing Myopia," *Harvard Business Review* 53 (No. 5, Sept.-Oct. 1975): 6.

11. A marketing agreement can be defined as a contract that obligates the producer to market his product with his cooperative and obligates the cooperative to accept the product. The agreement contains the rights and duties of both the cooperative member and the cooperative.

12. Some marketing agreements allow the farmer the option of transferring his commodity at the market price.

13. Carl E. Utterstrom, "Member Relations in Large and Complex Cooperatives" (M.A. thesis, University of Missouri, 1974).

14. William E. Black and James E. Haskell, "Vertical Integration through Ownership," *Marketing Alternatives for Agriculture: Is There a Better Way?* Committee on Agriculture and Forestry, U.S. Senate, 94th Congress, 2d Session, U.S. Government Printing Office, Washington, D.C., 1976, p. 54.

15. M. A. Adelman, "Integration and Antitrust Policy," 63HLR 27-58 (1949); A. R. Burno, *The Decline of Competition* (New York: McGraw-Hill, 1936).

16. A. R. Wyatt and D. E. Kieso, *Business Combinations: Planning—Action* (Scranton, Pa.: International Textbook Co., 1969).

17. Black and Haskell, "Vertical Integration through Ownership," p. 53.

18. For a discussion, see W. Smith Greig, "Regulation of Competition on Food Marketing," in Ronald D. Knutson and R. E. Schneidau, *The Economics of Food Processing* (Westport, Conn.: AVI Pulishing Co., 1971), Chap. 4.

19. Wyatt and Kieso, *Business Combinations,* p. 66.

20. Clement Ward, *Production-Marketing Alternatives for Cattlemen,* Special Report 21, FCS, USDA, Washington, D.C., Feb. 1976.

21. An excellent treatment of this subject is James D. Shaffer and Randall E. Torgerson, "Exclusive Agency Bargaining," *Marketing Alternatives for Agriculture,* pp. 38-48. Also see "Agricultural Marketing and Bargaining," Hearings before the Subcommittee on Domestic Marketing and Consumer Relations of the Committee on Agriculture, House of Representatives, 92d Congress, 1st Session, Sept. 20-Oct. 6, 1971.

22. H. R. 18706, 91st Congress, 2d Session, July 29, 1970.

23. *The American College Dictionary* (New York: Random House), p. 236.

24. Shaffer and Torgerson, "Exclusive Agency Bargaining," pp. 38-39.

25. Gilbert W. Biggs and J. Kenneth Samuels, *Bargaining Cooperatives: Selected Agri-Industries,* FCS Information 90, FCS, USDA, May 1973. For detailed study of bargaining, see Mahlon G. Lang, "Collective Bargaining between Producers and Handlers of Fruits and Vegetables for Processing: Setting Laws, Alternative Rules, and Selected Consequences" (Ph.D. dissertation, Michigan State University, 1977).

26. Charles E. French, "Statement before the Sub-Committee on Domestic Marketing and Consumer Relations," House Agriculture Committee, Sept. 1971, p. 11.

27. Ibid., p. 10.

28. Gene Ingalsbe, "In Michigan: Bargaining Law Extended; Having Its Day in Court," *News for Farmer Cooperatives,* FCS, USDA, Aug. 1976, pp. 8-10.

29. Jack H. Armstrong, "Institutions and Institutional Arrangements to Improve Producer Access to Markets," Working paper, FCS, USDA, 1975.

30. Dennis R. Henderson, Lee F. Schrader, and Michael S. Turner, "Centralized Remote-Access Markets," *Marketing Alternatives for Agriculture,* p. 9.

31. James V. Rhodes, Marketing Workshop, "Who Will Control the Swine Industry?" Speech to American Pork Congress, Indianapolis, March 14, 1973.

32. Olan D. Forker, *Price Determination Processes: Issues and Evaluation,* FCS Information 102, FCS, USDA, Sept. 1975, p. 26.
33. One option involving more effective enforcement of existing laws is outlined in Ronald D. Knutson, Dale C. Dahl, and Jack H. Armstrong, "Fine Tuning the Present System," *Marketing Alternatives for Agriculture,* pp. 82–90.
34. Henderson, Schrader, and Turner, "Centralized Remote-Access Markets," p. 13. For additional references on electronic exchanges, see David L. Holder, *A Computerized Forward Contract Market for Slaughter Hogs,* A.E.R. 211, Department of Agricultural Economics, Michigan State University, Jan. 1972; and Lee F. Schrader, Richard G. Heifner, and Henry E. Larzelere, *The Electronic Egg Exchange: An Alternative System for Trading Shell Eggs,* A.E.R. 119, Department of Agricultural Economics, Michigan State University, Dec. 1968.
35. Dennis R. Henderson, Lee F. Schrader, and Michael S. Turner, "Electronic Commodity Markets," Draft manuscript, Jan. 1975.
36. The auctioneer works *downward* from an initial asking price until the first bid is received. The product is then sold at the first bid.
37. Publically available USDA statistical information is discussed in *Major Statistical Series of the U.S. Department of Agriculture—How They Are Constructed and Used,* Vol. 1-11, Agriculture Handbook 365, 1970-72.
38. Market power can be defined as the ability to possess some degree of power over price.
39. David L. Holder, "One Hog Market for the U.S.," *National Hog Farmer* (St. Paul, Minn.: Webb Publishing Co., Oct. 1972).
40. Pricing efficiency is measured in terms of the accuracy, effectiveness, and speed of the price discovery process relative to cost.
41. There is the possibility that the auction exchange operators can have some influence over price by simply withholding offerings for limited time periods or by setting initial offering prices.

CHAPTER 5

1. Vertical integration might also be thought of as an organizational strategy. However, vertical integration is the major focus of the general strategy of integration and coordination and was discussed previously.
2. In this book, the word *merger* will be used to include consolidations and acquisitions.
3. Joseph G. Knapp, *Farmers in Business: Studies in Cooperative Enterprise,* American Institute of Cooperation, Washington, D.C., 1963, p. 269.
4. Bruce Swanson, "Pre- and Post-Merger Characteristics of Agricultural Cooperative Reorganizations and Implications for Planning, Financing, and Growth" (Ph.D. dissertation, Texas A & M University, College Station, Texas, Dec. 1975). Also Leon Garoyan and Gail L. Cramer, *Cooperative Mergers: Their Objectives, Success, and Impact on Growth,* Station Bulletin 605, Agricultural Experiment Station, Oregon State University, Feb. 1969.
5. Swanson, "Pre- and Post-Merger Characteristics," p. 326.
6. Ibid., p. 327.
7. Garoyan and Cramer, *Cooperative Mergers,* p. 21.
8. Ibid.
9. Swanson, "Pre- and Post-Merger Characteristics," p. 138.
10. Ibid., p. 52.
11. Willard F. Mueller, *The Role of Mergers in the Growth of Agricultural Cooperatives,* California Agricultural Experiment Station Bulletin 777, 1966, pp. 8, 10.
12. Swanson, "Pre- and Post-Merger Characteristics," p. 47.
13. Garoyan and Cramer, *Cooperative Mergers,* p. 22.
14. For an excellent source on this subject, see F. M. Scherer, *Industrial Market Structure and Economic Performance,* (Chicago: Rand McNally College Publishing Co., 1970), Chapter 20.
15. Occasionally such a merger will be blocked if the acquiring firm already has a significant market share.

16. For an indication of this and an analysis of the agricultural cooperative antitrust exemption, see the Federal Trade Commission, *Staff Report on Agricultural Cooperatives,* Sept. 1975.
17. Swanson, "Pre- and Post-Merger Characteristics," p. 52.
18. Of course, for many cooperatives internal expansion may be the most logical approach to horizontal growth.
19. H. Igor Ansoff, "Strategies for Diversification," *New Decision-Making Tools for Managers,* edited by Edward L. Bursk and John F. Chapman, The Mentor Executive Library, President and Fellows of Harvard College, 1963.
20. Philip Kotler, *Marketing Management, Analysis, Planning, and Control,* 3rd ed. (Englewood Cliffs, N.J.: Prentice-Hall, 1976), p. 49. Note the use of the conglomerate form of organizational structure as a subset of diversification.
21. Regulatory agencies are basically concerned with horizontal and to a lesser extent vertical types of merger growth. As a result, these agencies have come to consider conglomerate merger as any merger that cannot be classified as a pure horizontal or vertical merger.
22. Richard P. Rumelt, *Srategy, Structure, and Economic Performance* (Boston: Division of Research, Graduate School of Business Administration, Harvard University, 1974).
23. Albert G. Madsen and Richard G. Walsh, "Conglomerates: Economic Conduct and Performance," *Journal of Farm Economics* 51 (5, Dec. 1969): 1495.
24. Riceland Foods does sell some planting seed to farmers.
25. Sales of MFC Services were $200.6 million in fiscal year 1977. Farmland, Land O'Lakes, and Gold Kist had sales of $3.0 billion, $1.4 billion, and $1.1 billion respectively.
26. Ronald D. Knutson, "A Full Service Cooperative Concept: A Key to Leadership," Speech given at Farm Supply Cooperative Management Training Workshop, May 19, 1975.
27. Kotler, *Marketing Management, Analysis, Planning, and Control,* p. 51.
28. Ansoff, "Strategies for Diversification," p. 309.
29. For tax purposes cooperatives exempt under IRS Code Section 521 must keep net margins separate between marketing and supply divisions.
30. John D. Moorhead, "Gold Kist Climbs to Farm Cooperative Heights," *Christian Science Monitor,* Apr. 9, 1976.
31. Rumelt, *Strategy, Structure, and Economic Performance.*
32. "We'll Buy You Out," *Wall Street Journal,* May 1976.
33. The same provision is included in the Capper-Volstead Act.
34. Willard F. Mueller, "Firm Conglomeration as a Market Structure Variable," *Journal of Farm Economics* 51 (5, Dec. 1969): 1488.
35. John Kenneth Galbraith, *The New Industrial State* (New York: Houghton Mifflin Co., 1967).
36. Claud L. Scroggs, "Historical Highlights," *Agricultural Cooperation: Selected Readings,* edited by M. A. Abrahamsen and C. L. Scroggs (Minneapolis: University of Minnesota Press, 1957), p. 10.
37. Ernest V. Stevenson, "Positive Approach to Grain Marketing: Build on What We Already Have," *News for Farmer Cooperatives,* FCS, USDA, Washington, D.C., May 1975, p. 7.
38. *CF Industries, Inc., Annual Report 1975,* Long Grove, Ill.
39. *This is Universal Cooperatives, Inc.,* Pamphlet distributed by Universal Cooperatives, Inc., Alliance, Ohio.
40. *Universal Buyer-Merchandiser,* Albert Lea, Minn., 5 (6, Nov./Dec. 1977): 17.
41. *This Is the National Council of Farmer Cooperatives,* Pamphlet distributed by NCFC, Washington, D.C.
42. *This Is AIC: The National Educational Organization for Farmer Cooperatives,* Pamphlet distributed by AIC, Washington, D.C.
43. *Cooperatives, USA: Facts and Figures,* Cooperative League of USA, Washington, D.C., 1972.
44. The following reports are drawn on heavily in this section: Fred Hulse and Michael Phillips, *Joint Ventures Involving Cooperatives in Food Marketing,* Marketing Research Report 1040, FCS, USDA, May 1975; and Lester H. Meyers, Michael J. Phillips, and Ray A. Goldberg, "Joint Ventures Among Cooperative and Non-

cooperative Marketing Firms," *Marketing Alternatives for Agriculture: Is There a Better Way?* Committee on Agriculture and Forestry, U.S. Senate, 94th Congress, 2d Session, U.S. Government Printing Office, Washington, D.C., 1976.

45. Joseph G. Knapp, "Prospects for the Cooperative Form of Business Organization," Speech given at Graduate Institute of Cooperative Leadership, University of Missouri, July 24, 1973.

46. Ray A. Goldberg, "Profitable Partnerships: Industry and Farmer Co-ops," *Harvard Business Review* 50 (Mar.-Apr. 1972): pp. 108–21.

47. Usually cooperatives do not have a raw material supply problem. This eliminates the usual incentive found in cooperative-corporate joint ventures.

CHAPTER 6

1. Joseph G. Knapp, *Capper-Volstead Impact on Cooperative Structure,* FCS Information 97, FCS, USDA, Washington, D.C., Feb. 1975.

2. David Volkin, "Understanding Capper-Volstead," Reprint 392 from *News for Farmer Cooperatives,* FCS, USDA, Washington, D.C., July 1974.

3. The Capper-Volstead Act U.S.C. 291, 292; 42 Stat. 388 (1922) (Public—No. 146—67 Congress).

4. Joseph G. Knapp, "Are Cooperatives Good Business?" *Farmers in Business: Studies in Cooperative Enterprise,* American Institute of Cooperation, Washington, D.C., 1963, pp. 15–16.

5. George L. Baker, "Milk, Fruit, Feel the Stirrings of Antitrust," *New York Times,* Oct. 5, 1975, p. 7.

6. Linda Kravitz, *Who's Minding the Co-op?* Agribusiness Accountability Project, Washington, D.C., Mar. 1974.

7. Brooks Jackson, "Pressure by Farm Co-ops Blamed in Shelving of Criticism by GPO Unit," *Washington Post,* Oct. 5, 1974, p. A5.

8. Keith I. Clearwaters, "Challenges Facing Dairy Bargaining Cooperatives," Address to the Dairy Conference of the American Farm Bureau Federation's 56th Annual Meeting, New Orleans, La., Jan. 6, 1975, pp. 14–15.

9. Baker, "Milk, Fruit, Feel the Stirrings of Antitrust," p. 7.

10. Frank Lipson, Clint Batterton, and Alison Masson, *Federal Trade Commission: Staff Report on Agricultural Cooperatives,* U.S. Government Printing Office, Washington, D.C., Sept. 1975, p. 1.

11. Capper-Volstead Committee, "The Question of Undue Price Enhancement by Milk Cooperatives: A Summary," USDA, Washington, D.C., Dec. 1976, p. 1.

12. Ibid., p. 19.

13. Antitrust Immunities Task Force, "United States Department of Justice Report on Milk Marketing," Antitrust Division, U.S. Department of Justice, Washington, D.C., Jan. 17, 1977. Further discussion of this report is in the section under Marketing Orders.

14. USDA, "U.S. Department of Agriculture Comments on the Department of Justice Report on Milk Marketing," Washington, D.C., May 26, 1977, p. 7.

15. Executive Order 12022, *Federal Register,* Vol. 42, No. 233, Dec. 5, 1977.

16. USDA, "Testimony Submitted by the Honorable Bob Bergland, Secretary of Agriculture, to the National Commission for the Review of Antitrust Laws and Procedures," Washington, D.C., July 27, 1978, p. 10.

17. National Council of Farmer Cooperatives, "Policy Resolutions for Action by the Delegate Body," 49th Annual Meeting, Fairmont Hotel, San Francisco, Calif., Jan. 12, 1978.

18. J. K. Samuels, "Legal and Legislative Aspects," *Bargaining in Agriculture: Potentials and Pitfalls in Collective Action,* North Central Regional Extension Publication 30, University of Missouri Extension Division, C911, June 1971, p. 26.

19. Ibid., pp. 27 and 28.

20. Dan Hager, "Training Session Explores Use of Bargaining Act," *The Packer,* May 6, 1978, p. 19A.

21. Keith I. Clearwaters, "Challenges Facing Dairy Bargaining Cooperatives."

22. *Legal-Tax-Accounting Memorandum,* Vol. XXIV, No. 10, National Council of Farmer Cooperatives, June 19, 1974.
23. Alfred F. Dougherty, Jr., "Viewpoint: Federal Trade Commission," *How the Public Views Agricultural Bargaining,* Proceedings: 20th National Conference of Bargaining and Marketing Cooperatives, FCS Special Report 23, FCS, USDA, Washington, D.C., Jan. 1976, pp. 40–41.
24. Walter J. Armbruster, Truman F. Graf, and Alden C. Manchester, "Marketing Orders," *Marketing Alternatives for Agriculture: Is There a Better Way?* Committee on Agriculture and Forestry, U.S. Senate, 94th Congress, 2d Session, U.S. Government Printing Office, Washington, D.C., 1976, p. 66.
25. For milk marketing orders with individual handler pooling and for California citrus fruits, approval by at least three-fourths of the producers is required.
26. Alexander Swantz, *How Federal Milk Marketing Orders Are Developed and Amended,* USDA, Washington, D.C., Aug. 1952.
27. "Three important considerations need to be kept in mind. First, the above estimates apply to the share of the crops moving in regulated channels. In some instances the marketing order may have substantial influence on the share of the crop moving into alternative outlets (generally processing) not subject to regulation under the order program. Second, the estimates are based on crops *subject* to regulation, but not necessarily being regulated at the present time. Finally, the estimates cover only those crops to which orders apply; major crops such as apples, green peas, snap beans, and many others are not included." Walter J. Armbruster, "Fruit and Vegetable Marketing Orders," Presentation before the USDA Regulatory Advisory Committee Meeting, Washington, D.C., Oct. 20, 1976.
28. "Price Impacts of Federal Market Order Program," Report of the Interagency Task Force, Special Report 12, FCS, USDA, Washington, D.C., Jan. 7, 1975, pp. 29 and 44.
29. Ibid., p. 7.
30. USDA, *Federal Milk Order Market Statistics,* FMOS-221, Dairy Division, AMS, Washington, D.C., May 1978, p. 35.
31. Herbert L. Forest, "Goverment Regulation of Milk Marketing," *Proceedings: Cooperatives, Milk Marketing, and Regulations—A Symposium,* National Milk Producers Federation, Washington, D.C., Apr. 13 and 14, 1976, pp. 50–51.
32. For further discussion, see Armbruster, Graf, and Manchester, "Marketing Orders"; Glenn Lake, "Can Bargaining Succeed in the Dairy Industry Without Marketing Orders?" and Floyd F. Hedlund, "Can Bargaining Succeed in the Fruit and Vegetable Industry Without Marketing Orders?" both in *How the Public Views Agricultural Bargaining.*
33. Olan D. Forker and Brenda A. Anderson, *An Economic Evaluation of Methods Used Under Marketing Order Legislation,* A. E. Research 216, Department of Agricultural Economics, Cornell University Agricultural Experiment Station, Apr. 1976, p. 14.
34. National Council of Farmer Cooperatives, "Policy Resolutions—For Approval by the Delegate Body, 47th Annual Meeting," Washington, D.C., Jan. 15, 1976.
35. "Price Impacts of Federal Market Order Programs," p. 21.
36. *Economic Report of the President,* Transmitted to the Congress Feb. 1975, U.S. Government Printing Office, Washington, D.C., 1975, pp. 185–86.
37. Federal Trade Commission, *A Report on Agricultural Cooperatives,* Bureau of Competition, U.S. Government Printing Office, Washington, D.C., Sept. 1975.
38. Ibid., p. 178.
39. "USDA Comments on FTC Staff Report Relating to Cooperatives," *Congressional Record—Senate,* Feb. 19, 1976, p. S1958.
40. Ibid., p. S1962.
41. Aileen Gorman, "Agricultural Outlook for 1976: A Consumer Response," *1976 U.S. Agricultural Outlook,* Prepared for the Committee on Agriculture and Forestry, U.S. Senate, U.S. Government Printing Office, Washington, D.C., Dec. 18, 1975, p. 65.
42. Hedlund, "Can Bargaining Succeed in the Fruit and Vegetable Industry?" p. 29.
43. USDA, "Advisory Committee on Regulatory Programs," Secretary's Memorandum No. 1895, Mar. 5, 1976.
44. USDA, "Report on Activities of Advisory Committee on Regulatory Programs," Washington, D.C., July 22, 1977.

45. National Milk Producers Federation, *Proceedings: Cooperatives, Milk Marketing, and Regulations—A Symposium,* Washington, D.C., Apr. 13 and 14, 1976.

46. Arthur E. Rowse, "Consumer Contact—Food Rise Story Is Aired," *Baltimore Sun,* Baltimore, Md., June 3, 1976.

47. Antitrust Immunities Task Force, "United States Department of Justice Report on Milk Marketing," Antitrust Division, U.S. Department of Justice, Washington, D.C., Jan. 17, 1977.

48. Ibid., pp. xii.

49. Ibid., pp. 394–95.

50. Ibid.

51. USDA, "U.S. Department of Agriculture Comments on the Department of Justice Report on Milk Marketing."

52. Paul W. MacAvoy (ed.), *Federal Milk Marketing Orders and Price Supports,* American Enterprise Institute for Public Policy Research, Washington, D.C., Nov. 1977.

53. Martin E. Abel and Michele M. Veeman, "Marketing Boards," *Marketing Alternatives for Agriculture,* p. 73.

54. R. M. A. Loyns, "The Role and Impact of Marketing Boards in Agricultural and Food Stability," *Canadian Journal of Agricultural Economics,* Canadian Agricultural Economics Society, Workshop Proceedings, Mar. 1976, p. 69.

55. Ronald D. Knutson and Olan D. Forker, "The Options in Perspective," *Marketing Alternatives for Agriculture,* pp. 104–9.

56. For a detailed description and historical background, see *Legal Phases of Farmer Cooperatives: Federal Income Taxes,* No. 2, Farmer Cooperative Service, FCS Information 100, FCS, USDA, Washington, D.C., May 1976.

57. Lee Schrader and Ray Goldberg, *Farmers' Cooperatives and Federal Income Taxes* (Cambridge, Mass.: Ballinger Publishing Co., 1975), p. 17.

58. The requirement of 20 percent cash payout does not apply in the case of per unit retains which cooperatives are also allowed to deduct under Subchapter T.

59. Use of the term "exempt" is a misnomer since the status does not guarantee exemption from all income taxes. "Exempt" status refers only to the ability to deduct dividend payments and nonpatronage income. "Nonexempt" cooperatives cannot deduct these amounts, although they may still be able to deduct patronage dividends under Subchapter T.

60. R. K. Morris, "Comprehensive Review of Cooperative Tax Structure (including recent tax rulings)," *Cooperative Laws Are Different: Proceedings of the Second National Symposium on Cooperatives and the Law,* University of Wisconsin, Madison, Apr. 22–24, 1975.

61. The restriction exists from at least one other source as well. Borrowers from the Banks for Cooperatives must also satisfy the 50 percent rule.

62. Various experts have suggested that the IRS may be on somewhat shaky ground here since three basic federal cooperative statutes do not require compliance with the one-member, one-vote rule. These acts include the Capper-Volstead Act, the Agricultural Marketing Act of 1929, and the Farm Credit Act of 1971. Also, many state statutes permit voting on a basis other than one-member, one-vote.

63. Revenue Ruling 73–568, 1973–2.

64. Morris, "Comprehensive Review of Cooperative Tax Structure," p. 36.

65. Ibid., p. 35. Also see Jim Baarda, "Legal Corner," *Farmer Cooperatives,* ESCS, USDA, Washington, D.C., Mar. 1978.

66. Nelda Griffin, "Federal Income Tax Status of Farmer Cooperatives," *The Cooperative Accountant,* 31 (3, Fall 1978): 51.

67. Charles Kraenzle and Francis Yager, *Grain Marketing Patterns of Local Cooperatives,* Research Report 31, FCS, USDA, Oct. 1975.

68. Many cooperatives, upon giving up their exemption, switch to debt securities, thereby allowing deduction of interest payments.

69. See financial section for more details.

70. Lee Schrader, "Cooperatives and the Tax Law," Presentation at a workshop on Agricultural Cooperatives and the Public Interest, June 6–8, 1977, St. Louis, Mo.

71. National Council of Farmer Cooperatives, "Policy Resolutions for Approval by the Delegate Body," 47th Annual Meeting, Washington, D.C., Jan. 15, 1976.

72. For a description of how market news reports are constructed and used, see *Major*

Statistical Series of the U.S. Department of Agriculture: How They Are Constructed and Used, Vol. 10., Market News, Agriculture Handbook 365, 1972.

73. PCCA also markets raw cotton for American Cotton Growers, an integrated cooperative that operates a textile mill.

74. For a detailed discussion of this alternative see Kirby Moulton and Daniel I. Padberg, "Mandatory Public Reporting of Marketing Transactions," *Marketing Alternatives for Agriculture.* Another approach to improved government information involves fine tuning the present market information system. See Ronald D. Knutson, Dale C. Dahl, and Jack H. Armstrong, "Fine Tuning the Present System," *Marketing Alternatives for Agriculture,* pp. 87–88.

75. For an excellent source on promotion guidelines for agricultural groups, see Edward Dailey, *Guidelines—Advertising and Promotion of Farm Products,* Extension Circular 530, Cooperative Extension Service, Purdue University, West Lafayette, Ind., June 1964.

76. Carl R. Twining and Peter L. Henderson, *Promotional Activities of Agricultural Groups,* Marketing Research Report 742, ERS, USDA, Dec. 1965; data for 1973 are updates of this report, and findings reported here are preliminary and are not yet in published form. Includes expenditures for advertising, merchandising, public relations and consumer education, and promotional-related administrative costs.

77. The Consumer Price Index was used as the deflator.

78. Preliminary.

79. "Ad Spending by the Leading Farmer Co-ops," *Weekly Digest,* American Institute of Food Distribution, Inc., Dec. 1, 1973, p. 4.

80. Preliminary. Brand promotion can be defined as stimulating demand for a specific product brand to enlarge that brand's share of the market in competition with other brands of the same product. Nonbranded promotion, in the most general sense, is the stimulating of primary demand for a product to enlarge total demand in competition with other similar products.

81. Cotton Incorporated also receives funding from Congress for research. However, that source of funds has been diminishing.

82. *Cotton and Wool Situation,* CWS-4, ERS, USDA, Washington, D.C., Mar. 1976, p. 6.

83. William W. Gallimore, *Synthetics and Substitutes for Agricultural Products: Projections for 1980,* Marketing Research Report 947, ERS, USDA, Washington, D.C., Mar. 1972.

84. Gallimore, *Synthetics and Substitutes,* p. ii.

85. For a good background report on the subject, see *Edible Soy Protein: Operational Aspects of Producing and Marketing,* Research Report 33, FCS, USDA, Washington, D.C., Jan. 1976.

86. Gallimore, *Synthetics and Substitutes,* p. 24.

87. USDA, *U.S. Foreign Agricultural Trade Statistical Report, Calendar Year 1976,* ERS, USDA, Washington, D.C., June 1977, p. 1.

88. Henry W. Bradford and Richard S. Berberich, *Foreign Trade of Cooperatives,* FCS Information 88, FCS, USDA, U.S. Government Printing Office, Washington, D.C., Feb. 1973, p. 5.

89. Donald E. Hirsch, "Agricultural Exports by Cooperatives, 1976," ESCS, USDA, Unpublished manuscript. "Exports were classifed as 'direct exports' when cooperatives themselves sold the commodities overseas. They were classified as 'indirect sales' when cooperatives sold commodities to other U.S. organizations which took title to the commodities and negotiated overseas sales." For more technical information on exporting, see Donald E. Hirsch, *Export Techniques of Grain Cooperatives,* FCS Information 104, FCS, USDA, Washington, D.C., Apr. 1976, 45 pp.

90. Stanley K. Thurston, Michael J. Phillips, James E. Haskell, and David Volkin, *Improving the Export Capability of Grain Cooperatives,* FCS Research Report 34, FCS, USDA, Washington, D.C., June 1976; and Donald E. Hirsch, "Cooperatives Directly Export $2 Billion in Farm Products," *Farmer Cooperatives,* ESCS, USDA, Washington, D.C., May 1978, p. 9.

91. Jack Frost, "Amcot: New Look in Marketing Cotton," *Blueprint for Cooperative Growth, Action, Innovation, Coordination,* American Institute of Cooperation, Washington, D.C., 1972–73, pp. 221–25.

92. The largest 100 agricultural cooperatives are defined as those with the highest gross sales figures provided they also had assets of at least $5 million. Nelda Griffin, *A Financial Profile of Farmer Cooperatives in the United States,* FCS Research Report 23, FCS,

USDA, Washington, D.C., Oct. 1972; and Nelda Griffin, "Trends in Cooperative Capitalization and Financial Structure for the 100 Largest U.S. Farmer Marketing and Supply Cooperatives," Presentation at the National Council of Farmer Cooperatives' annual meeting, Jan. 10, 1978.

93. Philip Brown and David Volkin, *Equity Redemption Practices of Agricultural Cooperatives,* FCS Research Report 41, USDA, Washington, D.C., Apr. 1977, pp. 4, 9.
94. Taken from "This Is AIC," American Institute of Cooperation.
95. W. M. Harding, "Current and Anticipated Issues in Financing Farm Supply and Marketing Businesses," Presentation to the Graduate Institute of Cooperative Leadership, Columbia, Mo., July 24, 1975.
96. Wilmer A. Dahl and W. D. Dobson recently completed a study that accounts for the farmer opportunity cost of cooperative revolving fund capital. Conclusions indicated that a least-cost financial structure for Wisconsin farm supply cooperatives should include more permanent dividend-paying equity capital (stock), more debt, and substantially less revolving fund capital. "An Analysis of Alternative Financing Strategies and Equity Retirement Plans for Farm Supply Cooperatives," *American Journal of Agricultural Economics* 58 (2, May 1976): 198.
97. Ibid. Also see A. R. Tubbs, "Capital Investments in Agricultural Marketing Cooperatives: Implications for Farm Firm and Cooperative Finance" (Ph.D. dissertation, Cornell University, 1971).
98. Brown and Volkin, *Equity Redemption Practices,* pp. 21, 25.
99. In fact, several marketing cooperative managers pointed out that their low equity positions would make it impossible to borrow money from any source other than the Banks for Cooperatives. See also John Moore and Richard Fenwick, "Future Cooperative Growth-Capital Requirements," *American Institute of Cooperation Yearbook, 1977-78,* p. 245.
100. Growers put up an average of $14,000 each.
101. Includes capital leases, industrial revenue bonds, other agricultural cooperatives, insurance companies, noncooperative businesses, national and state farm organizations, credit unions, and employee trust funds.
102. The report was headed by an official of the FCA and sixteen others from the banking and accounting industries, the universities, and cooperative management. *Report of the Task Force on Cooperative Finance,* Farm Credit Administration, Oct. 1, 1975.
103. Ibid., p. 9.
104. Ibid., p. 10.
105. Ninety percent of Agway's recent $30 million issue of debentures was sold to Agway members. Other cooperatives have also had highly successful sales of securities to members.
106. Bernard H. Schulte, "How Much Equity Capital Is Needed?" Paper presented at the 1974 meeting of the American Institute of Cooperation, Kansas State University, Manhattan.
107. For further information see Donald Davidson, "Cooperatives Report Increasing Use of Industrial Development Bonds," *Farmer Cooperatives,* June 1978, p. 7.
108. Ibid., p. 4.
109. For further information see *Leveraged Leasing for Cooperatives (An Innovative Technique for Financing Large-Scale Projects),* the New Orleans Bank for Cooperatives, New Orleans, La., in cooperation with the Louisville Bank for Cooperatives, Louisville, Ky., Apr. 1974; also Peter Vanderwicken, "The Powerful Logic of the Leasing Boom," *Fortune,* Nov. 1973, p. 132.
110. "Tax Brief," *Business Week,* Sept. 1, 1975.
111. The SEC uses the Internal Revenue Code qualifications for the Section 521 status to serve as qualification for exemption from SEC registration requirements. As a result, once a cooperative gives up its Section 521 status it also gives up its special exemption for SEC securities registration.
112. The federal law further details these exemptions along with certain exceptions.
113. Jerome Weiss, "So You Think You're Exempt from Federal Securities Laws," *News for Farmer Cooperatives,* Apr. 1975, pp. 6-9.
114. Nelda Griffin, "Securities Situation Concerning Farmer Cooperatives," *News for Farmer Cooperatives,* Apr. 1976, p. 4.
115. Ibid., pp. 5-6.

CHAPTER 7

1. Peter F. Drucker, *Management: Tasks, Responsibilities, Practices* (New York: Harper & Row, 1974), p. 125.
2. Richard W. Schermerhorn, "Feasibility Analysis: An Integral Part of Growth." A. E. Series 102, Paper presented at the Annual Meeting of the Idaho Cooperative Council, Idaho Falls, Nov. 1971, p. 2.
3. D. W. Krager and A. Malik, "Long-Range Planning and Organizational Performance," *Long-Range Planning* 3 (6, Dec. 1975). An earlier article by S. S. Thune and R. J. House, "Where Long-Range Planning Pays Off," *Business Horizons,* Aug. 1970, presents similar results.
4. While the term *long-range* serves to emphasize the time dimension of this activity, the more important feature is that it is total planning.
5. Two good articles showing the integration of the different levels of planning are J. R. Champion, "Corporate Planning in CPC Europe," *Long-Range Planning* (Dec. 1970): 8; and Basil W. Denning, "Introduction," *Corporate Planning: Selected Concepts* (London: McGraw-Hill, 1971), pp. 1–35.
6. A good example of a company that has a commitment to the future is Sun Oil Co. (*Business Week,* Nov. 8, 1976, p. 72). In light of the specter of no economically recoverable oil in the next century, it has already started to "de-integrate" its oil business as a prelude to diversification.
7. P. E. Holden, L. S. Fish, and H. L. Smith, *Top Management Organization and Control* (New York: McGraw-Hill, 1941), pp. 4–5.
8. A. D. Baker, Jr., and G. C. Thompson, "Long-Range Planning Pays Off," *Conference Board Business Record,* Oct. 1956, pp. 435–43.
9. "How 101 Companies Handle Corporate Planning," *Business Management,* Sept. 1967, p. 24.
10. Robert M. Fulmer and Leslie W. Rue, *The Practice and Profitability of Long-Range Planning* (Oxford, Ohio: Planning Executives Institute, 1973), p. 1.
11. The term *headquarters* is used here for the sake of efficiency to indicate varying combinations of board and management effort.
12. Directors most suitable for this assignment are those who have some familiarity with the business, have some planning experience, and are in a position to devote time and thought to the effort.
13. H. Steiglitz, *The Chief Executive—And His Job* (New York: National Industrial Conference Board, 1969), p. 21.
14. "Profile of a President," Heidrick and Struggles, Inc., Management Consulting and Executive Selection, Boston, 1972, p. 4.
15. Patrick Irwin, *Business Planning—Key to Profit Growth* (Toronto: Ryerson Press, 1969), p. 24.
16. Leslie W. Rue, "The How and Who of Long-Range Planning," *Business Horizons* 16 (6, Dec. 1973).
17. Throughout this section of the study, comparisons of noncooperative and cooperative planning were made using the study by Fulmer and Rue, *Practice and Profitability of Long-Range Planning.*
18. Based on data collected for a report by Gilbert W. Biggs, *Farmer Cooperative Directors: Characteristics, Attitudes,* FCS Research Report 44, ESCS, USDA, Feb. 1978.
19. K. Bandy, "FCX Designs New Planning Process," *News for Farmer Cooperatives,* FCS, USDA, Mar. 1971.
20. Richard L. Larson, *1975 Chief Executive Study,* Report to National Council of Farmer Cooperatives, Kearney: Management Consultants, San Francisco, Calif.
21. Leon Garoyan and Paul O. Mohn, *The Board of Directors of Cooperatives,* Division of Agricultural Sciences, University of California, July 1976, p. 23. The statement could prove fallacious for some cooperatives. For example, the manager of one cooperative told researchers that when they decided to start using formal planning, managers and directors sat down together to discuss their business. Only then did management realize that the directors knew a lot about their industry but virtually nothing about the company's markets.
22. Conference Board, *Corporate Directorship Practices—Membership and Committees of the Board,* Report 588, New York, 1973.

23. M. S. Nicolson, *Duties and Liabilites of Corporate Officers and Directors* (Englewood Cliffs, N.J.: Prentice-Hall, 1972).
24. *Profile of the Board of Directors,* Heidrick and Struggles Management Consulting, Chicago, 1971, p. 4.
25. Biggs, *Farmer Cooperative Directors,* pp. 39–40.
26. Eric Thor, a former administrator of the Farmer Cooperative Service, USDA, suggests that the ideal size for a cooperative board is 20 to 25 directors with an executive committee of 7 to 8 members.
27. *Profile of the Board of Directors,* Heidrick and Struggles, p. 5.
28. Biggs, *Farmer Cooperative Directors,* p. 23.
29. Based on data collected for a report by Biggs, *Farmer Cooperative Directors.*
30. *Profile of the Board of Directors,* Heidrick and Struggles, p. 6.
31. Prentice-Hall Editorial Staff, *Directors' and Officers' Encyclopedic Manual* (Englewood Cliffs, N.J.: Prentice-Hall, 1975), p. 142.
32. Biggs, *Farmer Cooperative Directors,* p. 6.

CHAPTER 8

1. Gilbert W. Biggs, *Farmer Cooperative Directors: Characteristics, Attitudes,* FCS Research Report 44, ESCS, USDA, Washington, D.C., Feb. 1978.
2. Competition as used here refers to a conscious striving against other business firms for customers.

INDEX